AF363304

Cast Irons

Cast Irons

Properties and Applications

Editor

Paolo Ferro

MDPI • Basel • Beijing • Wuhan • Barcelona • Belgrade • Manchester • Tokyo • Cluj • Tianjin

Editor
Paolo Ferro
University of Padua
Italy

Editorial Office
MDPI
St. Alban-Anlage 66
4052 Basel, Switzerland

This is a reprint of articles from the Special Issue published online in the open access journal *Metals* (ISSN 2075-4701) (available at: https://www.mdpi.com/journal/metals/special_issues/cast_irons).

For citation purposes, cite each article independently as indicated on the article page online and as indicated below:

LastName, A.A.; LastName, B.B.; LastName, C.C. Article Title. *Journal Name* **Year**, *Article Number*, Page Range.

ISBN 978-3-03943-142-7 (Hbk)
ISBN 978-3-03943-143-4 (PDF)

Contents

About the Editor

Paolo Ferro (Prof. Dr.) is currently an associated professor of Metallurgy and Materials Selection at the University of Padova (Italy). After obtaining his degree in Materials Engineering (Summa Cum Laude), he completed his PhD at the University of Padova in Metallurgical Engineering. He was the scientific director of the research program 'Numerical and Experimental Determination of Residual Stresses in Welded Joints and their Influence on Fatigue Strength' (Young Researchers Project, 2003–2004). He won the prize for young researchers 'Aldo Daccò' 2002. He is a member of CMBM (Centre for Mechanics of Biological Materials), coordinator of the European project on critical raw materials named DERMAP (Design of Components in a Critical Raw Material Perspective), member of the presidential council of the Italian Group of Fracture (IGF) and scientific coordinator of the Italian research program on the mechanical characterization of heavy section ductile irons. His research is mainly focused on (1) the analytical and numerical modelling of welding and heat treatment processes, (2) local criteria based on notch stress intensity factor (NSIF) and strain energy density (SED) for the fatigue strength assessment of pre-stressed components; (3) modelling of intermetallic phases evolution during heat treatments of duplex and superduplex stainless steels; (4) mechanical and metallurgical characterization of cast irons with particular attention to heavy section ductile iron castings and solution strengthened ferritic ductile irons (SS-FDI). He is the author of several papers and the editor of a book titled Residual Stress Analysis On Welded Joints by Means of Numerical Simulation and Experiments (IntechOpen Publishing).

Preface to "Cast Irons"

As a researcher and materials engineer, I have dedicated all my professional life, from 2000 to today, to the study of the intrinsic correlation between materials, process parameters and properties, since I believe that it is the key to better understanding the behaviour of materials, and the basis of excellent mechanical design. Due to their great technological importance, a significant part of my research has been focused on the study of cast irons. In particular, in recent years, I have focused on the mechanical characterization of heavy section ductile irons and new generation cast irons, to which I have also dedicated a research doctorate. Thanks to the journal Metals, I have had the privilege of managing the Special Issue 'Cast Irons: Properties and Applications', that is aimed at deepening the material–process–properties correlation of this extremely important family of alloys. The good news is that, right now, that Special Issue is becoming a book. The topics covered in the various chapters range from the study of technological properties, such as wear resistance and weldability, to both static and fatigue mechanical properties. Each chapter was written by excellent professionals and experienced researchers in the field. The book is addressed to PhD students, engineers, designers, researchers and professionals who need to deepen the above-mentioned important material–process–properties correlation applied to cast iron of different grades. I hope that the reader could appreciate this book and find in it what he is looking for.

Paolo Ferro
Editor

Article

The Role of Microstructure on Tensile Plastic Behavior of Ductile Iron GJS 400 Produced through Different Cooling Rates, Part I: Microstructure

Giuliano Angella [1], Dario Ripamonti [1,*], Marcin Górny [2], Stefano Masaggia [3] and Franco Zanardi [3]

[1]   Institute of Condensed Matter Chemistry and Technologies for Energy (ICMATE), National Research Council of Italy, via R. Cozzi 53, 20125 Milano, Italy; giuliano.angella@cnr.it
[2]   Faculty of Foundry Engineering, Department of Cast Alloys and Composites Engineering, AGH University of Science and Technology, 30-059 Kraków, Poland; mgorny@agh.edu.pl
[3]   Zanardi Fonderie S.p.A., via Nazionale 3, 37046 Minerbe (VR), Italy; MST@zanardifonderie.com (S.M.); franco.zanardi@icloud.com (F.Z.)
*   Correspondence: dario.ripamonti@cnr.it; Tel.: +39-02-66173-323

Received: 29 August 2019; Accepted: 26 November 2019; Published: 29 November 2019

Abstract: A series of samples made of ductile iron GJS 400 was cast with different cooling rates, and their microstructural features were investigated. Quantitative metallography analyses compliant with ASTM E2567-16a and ASTM E112-13 standards were performed in order to describe graphite nodules and ferritic grains. The occurrence of pearlite was associated to segregations described through Energy Dispersive X-ray Spectroscopy (EDS) analyses. Results were related to cooling rates, which were simulated through MAGMASOFT software. This microstructural characterization, which provides the basis for the description and modeling of the tensile properties of GJS 400 alloy, subject of a second part of this investigation, highlights that higher cooling rates refines microstructural features, such as graphite nodule count and average ferritic grain size.

Keywords: ductile iron; cooling rate; segregation; microstructure

---

1. Introduction

Ductile Irons (DIs) are ternary Fe-C-Si alloys in which graphite forms as spheroidal particles (nodules), allowing for a good compromise between mechanical properties and a low production cost [1,2]. The number of graphite nodules and their shape are the result of a various technological factors which influence cooling rate and physicochemical state of the liquid metal [1,3,4]. The cooling rate is mainly affected by the wall thickness, the thickness of the neighboring parts of the casting section, and the initial temperature of the metal and mold and the mold material to absorb heat. The physicochemical state of the liquid metal is in turn affected primarily by the chemical composition, charge materials, furnace atmosphere, holding time, liquid metal superheating, preconditioning, spheroidization, and inoculation processes used in the foundry practice [1–11]. The cooling conditions under which the eutectoid reaction takes place together with alloying elements influence the metallic matrix microstructure [12–14]. So, the production route to design and shape optimal ductile iron microstructures with proper mechanical properties is very complex, involving aforementioned different factors, as well as implemented heat treatment conditions [15,16].

Silicon is a graphitizer element which hinders the occurrence of iron carbide. Its effect is estimated via the CE (Carbon Equivalent) relationship, CE = %C + 1/3%Si. A CE value of 4.26 denotes the eutectic composition [1]. Silicon seems to play a negligible role in determining the ferrite grain size, and it can segregate around the graphite nodules, thus being a possible cracking site [17].

Copper is a common alloying element in DI because of its graphitizing effect. It promotes pearlite formation, in particular, when coupled with small Mn additions [18].

The chemical composition together with the cooling conditions after casting affect the microstructure of the alloy. A number of parameters, describing the cooling curve, can be found in literature (a list is provided in [1,2]) that may be related to the graphite shape. In this work, the transformation temperatures and the corresponding undercooling will be taken into account. The inoculation practice and the cooling rate cooperates to control the nodule count, while the conditions under which the eutectoid reaction takes place influence the matrix microstructure [12,13].

The tensile plastic behavior of ductile iron is very sensitive to microstructure and casting defects. In this connection, strain hardening analysis is a powerful tool to study the effect of microstructure on its tensile plastic behavior of ductile iron. Angella et al. [19] shows that the dislocation-density-related Voce equation describes properly the correlations between strain hardening and microstructure of metallic alloys. From published literature [19–24], there is limited information on the effect of microstructure on tensile plastic behavior of ductile iron in terms of the strain hardening effect and micro-mechanisms occurring during deformation of its microstructure. Hence, the tensile flow curves modeling associated with an explicit correlation between plastic behavior and some microstructure parameters have not yet been clearly disclosed. This work, which provides the microstructural basis for the description and modeling of the tensile behavior of GJS 400 alloy [25], will investigate the correlation between the cooling rates near eutectic and eutectoid transformations and the microstructural features of the alloy. Cooling rates are estimated through MAGMAsoft v.5.3 taking into account the solidification of actual samples.

2. Materials and Methods

The chemical composition of the GJS 400 produced by Zanardi Fonderie S.p.A. (Minerbe-VR, Italy) is reported in Table 1. Carbon and sulfur contents were measured through a combustion infrared detection technique with a LECO CS744 by LECO (St. Joseph, MI, USA), while the other elements were detected by optical emission spectrometer with a ARL3460 by Thermo Fisher Scientific (Waltham, MA, USA). The value of CE is 4.45%, which makes the alloy hypereutectic. The residual Mg is 0.046%, which allows for graphite spheroidization [2].

Table 1. Chemical compositions of GJS 400 alloy (wt%).

C	Si	Mn	Cu	Ni	Cr	Mg	P	S	Fe
3.63	2.45	0.129	0.133	0.0168	0.023	0.046	0.038	0.0061	Bal.

Nodularization treatment was performed in a tundish cover ladle, using a Fe–Si–Mg alloy (Si 45 wt%, Mg 6,5 wt%), together with the alloying elements needed to achieve the desired chemical composition. After alloying, the melt was gravity poured in horizontal green sand molds (silica sand with clay and sea–coal addition, plus 3.5% water to activate clay), shaped with a pattern plate and formed with a green sand molding plant, in order to obtain the following samples complying with EN 1563 standard [26], namely (Figure 1):

1. a Lynchburg sample with 25 mm diameter; and
2. three Y-blocks samples with thickness 25, 50, and 75 mm.

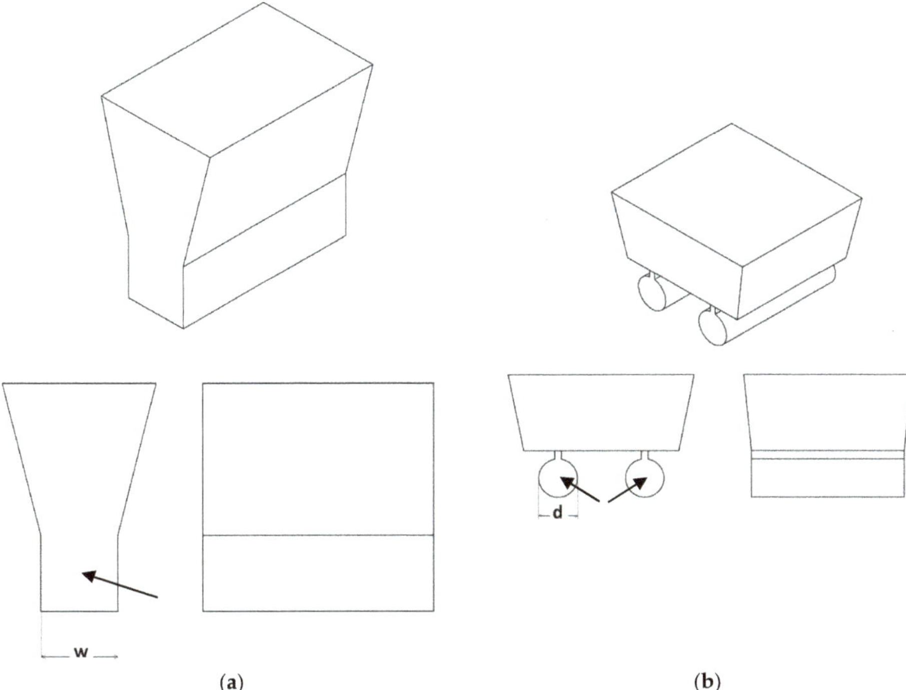

Figure 1. Sketches of the samples used. The upper part "feeds" the lower one, at the barycenter of which specimens were taken (see arrows). (**a**) 3D representation and orthographic projections of Y-block sample, w = 25, 50, 70 mm; (**b**) 3D representation and orthographic projections of Lynchburg sample, d = 25 mm.

The liquid metal was poured into the molds through the pouring basin and then, by mean of the gating system, it filled the cavity of all the samples. Specimen for metallographic analyses were taken in the lower part of the samples (see Figure 1). In particular, six specimens from each samples were investigated through Scanning Electron Microscopy (SEM) with a SU70 microscope by Hitachi (Tokyo, Japan) equipped with Energy Dispersive X-ray Spectroscopy (EDS) detector (Noran 6 system by Thermo Fisher Scientific (Waltham, MA, USA) for elemental microanalysis. The acceleration voltage was 20 kV and the working distance was about 15 mm. After conventional mechanical polishing, the samples were etched with Nital 10% for 5 s to highlight the grain boundaries of the ferritic matrix and the pearlitic islands. Nodule count, nodularity, average diameter of the graphitic nodules, and volumetric fractions of graphite and pearlite were determined through Digital Image Analysis, by means of ImageJ software [27], of SEM images complying with ASTM standard E2567-16a [28], whilst the determination of the average ferritic grain size was carried out through OM complying with ASTM standard E112-13 [29]. ASTM standards were chosen because to the authors' experience they are more commonly used.

ASTM standard E2567-16a requires that at least 500 graphite particles with a minimum MFD (Maximum Feret diameter) of 10 μm must be analyzed. A particle with a shape factor (ratio between the area of the particles and the area of the reference circle, this latter being related to the MFD) higher than 0.60 is defined as a nodule. Nodularity is then defined as the ratio between the total area of the nodules and the total area of the graphite particles. Nodule count is given by the ratio of the nodules and the test area, expressed in mm^2.

Grain size measurement were performed through the Hilliard single-circle procedure described in the ASTM standard E112-13. A single circle was blindly applied on at least five fields. A minimum of 35 intercepts between the circle and the grain boundaries is required. The ASTM grains size G is calculated as a function of the mean intercept, i.e., the ratio of the test line and the number of intercept. The average grain size can be thus calculated.

Since no direct measurement was possible, simulations of temperatures during cooling were performed through the Iron Module of the commercial software MAGMASOFT v5.3 by MAGMA (Aachen, Germany) in order to correlate cooling conditions with the microstructure. The inputs for this simulation are the 3D geometry of the casting system, the chemical composition of the alloy, the thermophysical parameters of the materials involved and alloy-mold and mold-environment heat transfer coefficients. The thermophysical parameters of the green sand, in particular the thermal conductivity, used for the simulation were determined by Zanardi Fonderie S.p.A. through an extensive experimental campaign aimed at the fine tuning of the parameters governing the heat fluxes. The actual set up of the gravity casting process was taken into account.

3. Results

3.1. Simulated Cooling Curves

The molten metal experienced significantly different solidification rates. Simulations of the casting system consisting of molten metal poured into sand molds were performed in Zanardi Fonderie S.p.A., and the cooling curves are reported in Figure 2. Data refer to the barycenter of the lower portion of the samples, where the specimens for metallographic analyses were taken. Eutectic (T_s) and eutectoid (T_e) equilibrium temperatures can be estimated on the basis of the chemical composition [14,30]:

$$T_s = 1154°C + 5.25\%Si - 14.88\%P = 1166.3 \ °C; \tag{1}$$

$$T_e = 739°C + 18.4\%Si + 2\%Si^2 - 14\%Cu - 45\%Mn + 2\%Mo - 24\%Cr - 27.5\%Ni + 7.1\%Sb = 787.8, °C \tag{2}$$

where "%el" represents the weight content of the element in the alloy. These equations hold for Si content up to 3 wt%, Mn, Cu, Cr, Ni content up to 1 wt%, and Mo content up to 0.5 wt% [14].

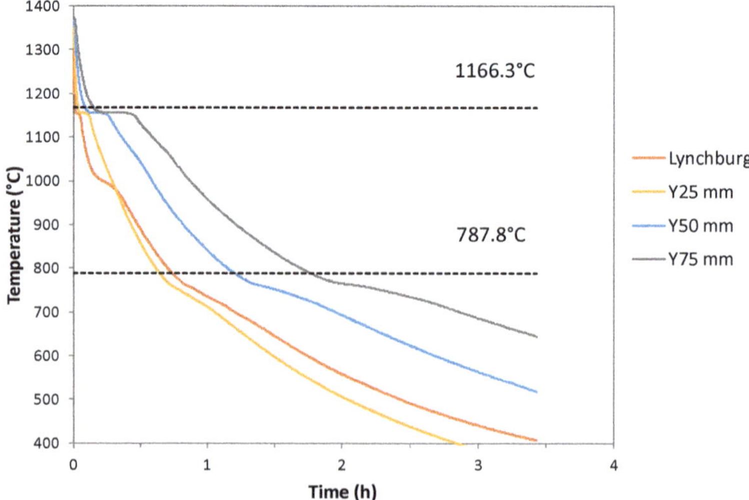

Figure 2. Simulations of temperature versus time of GJS 400 for the four different samples' geometry. The two dotted black lines represent eutectic and eutectoid equilibrium temperatures calculated through Equations (1) and (2), 1166.3 and 787.8 °C, respectively.

It can be seen (Figure 2 and Figure 4) that in the neighborhood of transformation temperatures the slope of the cooling curves varies abruptly because of the exothermic nature of eutectic and eutectoid transformation upon cooling. It can also be seen that for the Lynchburg sample at about 1000 °C the alloy experiences a reduction in cooling rate, which is an effect of the solidification occurring in the feeder.

As shown in Figure 3, indeed, the temperature decreases slower when the metal in the feeder undergoes solidification, an effect that disappears once solidification is complete. This phenomenon is not apparent in other molds because of their different geometries, and it is thought that it does not affect significantly microstructural features because it occurs far from the transformation temperatures.

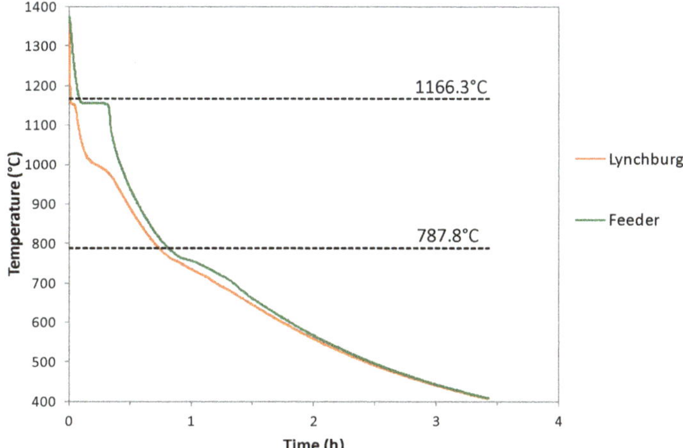

Figure 3. Simulations of temperature versus time of GJS 400 in different portions of the Lynchburg sample. When the alloy in the feeder undergoes solidification, cooling in the alloy in the lower portion is reduced. The two dotted black lines represent eutectic and eutectoid equilibrium temperatures calculated through Equations (1) and (2), 1166.3 and 787.8 °C, respectively.

Cooling rates near T_s and T_e (eutectic and eutectoid equilibrium temperatures, respectively) are given in Figure 4a,b, respectively. Table 2 summarizes cooling rates at the transformation temperatures, together with the undercooling experienced by the four samples, calculated as the difference between the eutectic temperature according to Equation (1) and the minimum temperature at the beginning of solidification.

Table 2. Undercooling at the eutectic transformation and cooling rate at transformation points for the four samples. Undercooling is calculated as the difference between the eutectic temperature according to Equation (1) and the minimum temperature at the beginning of solidification.

Mould	Undercooling (°C)	Cooling Rate at T_s [1] (°C/s)	Cooling Rate at T_e [2]
Lynchburg	11.56	1.98	0.09
Y25mm	11.39	0.56	0.11
Y50mm	10.45	0.16	0.06
Y75mm	9.96	0.10	0.04

[1] 1166.3 °C, according to Equation (1); [2] 787.8 °C, according to Equation (2).

Figure 4 and Table 2 show that the Lynchburg sample provided the fastest solidification rate, while at the eutectoid temperature the cooling rate is the second highest. It is worth noting that the differences in cooling rates are much higher at the eutectic temperature (there is a factor of about 20 between the highest and the lowest cooling rate), while at the eutectoid temperature they are comparable (only a factor of about 3). Moreover, variations in cooling rates are much higher in the proximity of the eutectic temperature rather than around eutectoid temperature, as an effect of reduced heat transfer from metal to heated mold.

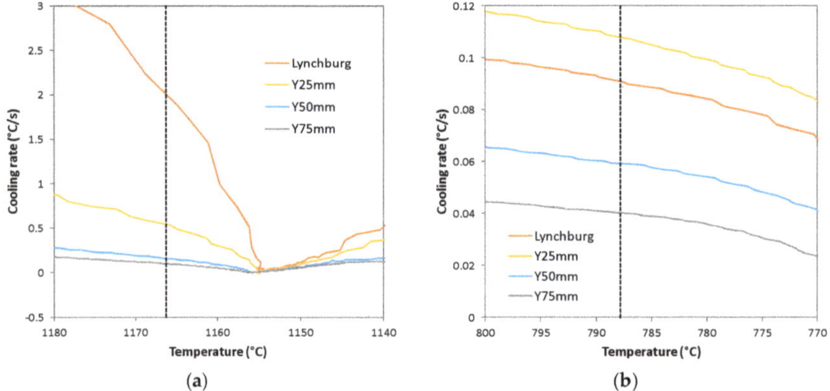

Figure 4. Cooling rate near to the equilibrium transformation temperatures calculated through Equations (1) and (2) for the four samples: (**a**) next to the eutectic temperature T_s, 1166.3 °C, calculated according to Equation (1) and indicated by the dotted black line (**b**) next to the eutectoid temperature T_e, 787.8 °C, calculated according to Equation (2) and indicated by the dotted black line. Steps are due to numerical derivation.

3.2. Microstructure

In Figure 5, representative SEM micrographs from Secondary Electron Imaging (SEI) of GJS 400 produced from the four different samples are reported. With slower solidification rates (Figure 4a) the microstructure became apparently coarser, with an evident increase of nodule size, while pearlite was present only in the specimens from Y-block samples (Figure 5b–d), and barely detectable in the specimens from Lynchburg sample (Figure 5a). This qualitative description can be supported through quantitative measurements according to ASTM standard E2567-16a [16]. Table 3 presents the results of image analysis, showing measurements on graphite features, defined in Section 2, and calculations on the volume fractions of the constituents. Together with the mean values, individual values measured on each specimen from each of the four samples are given.

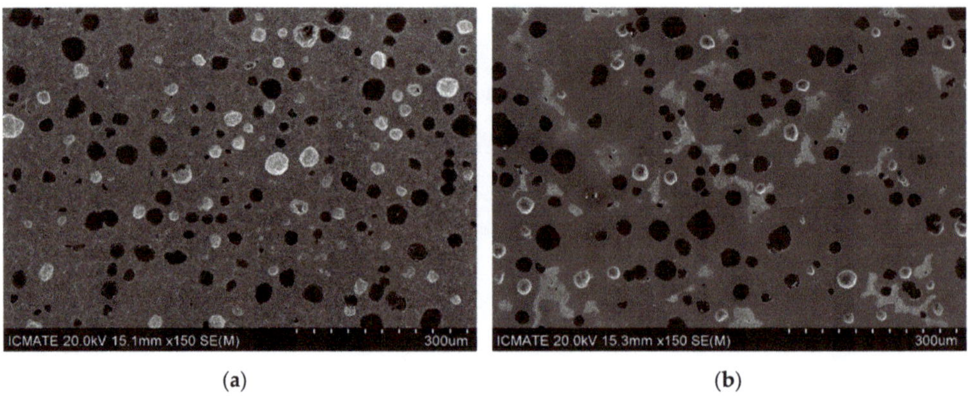

(**a**)   (**b**)

Figure 5. *Cont.*

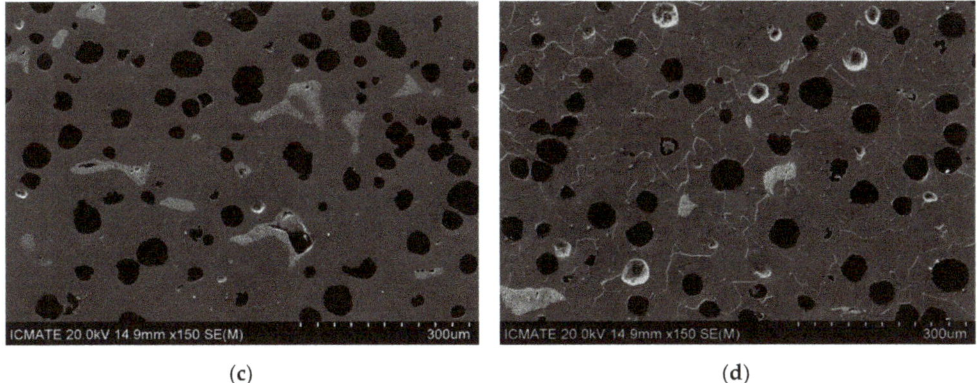

Figure 5. SEM micrographs (SEI) of GJS 400 produced through four different samples; (**a**) Lynchburg; (**b**) Y 25 mm; (**c**) Y 50 mm; (**d**) Y 75 mm. Pearlitic islands are present only in Y-block samples.

Table 3. Image analysis results for the specimens from the four samples.

Sample	Specimen	Graphite Features			Volume Fractions			Ferrite Grain Size (μm)
		Nodule Count (1/mm^2)	Nodularity (%)	Mean Diameter (μm)	Graphite (%)	Ferrite (%)	Pearlite (%)	
Lynchburg	1	241	85.7	24.4	13.6	86.4	-	38.7
	2	256	86.5	23.9	13.2	86.7	-	34.2
	3	285	90.9	23.6	13.8	86.0	-	39.4
	4	254	92.1	25.2	14.0	85.8	-	40.8
	5	261	92.8	24.6	13.5	86.5	-	32.5
	6	268	90.8	24.1	13.5	86.2	-	38.0
	Mean	261 ± 15	89.8 ± 3.0	24.3 ± 0.6	13.6 ± 0.3	86.3 ± 0.4	-	37.3 ± 3.0
Y 25 mm	1	242	91.4	24.5	12.9	83.1	4.1	43.1
	2	233	92.5	25.4	13.1	83.0	3.9	38.9
	3	255	92.9	25.2	13.9	82.6	3.5	38.1
	4	227	88.9	24.2	11.8	85.0	3.2	40.4
	5	240	89.7	25.4	13.6	82.2	4.2	38.1
	6	253	91.5	24.4	12.9	83.0	4.1	36.7
	Mean	242 ± 11	91.2 ± 1.6	24.9 ± 0.5	13.0 ± 0.7	83.1 ± 1.0	3.9 ± 0.4	39.2 ± 2.3
Y 50 mm	1	139	88.8	30.6	11.9	84.9	3.2	50.3
	2	117	85.1	30.0	10.5	86.4	3.1	41.6
	3	95	85.8	32.6	10.0	84.4	5.6	46.2
	4	119	87.0	31.7	11.3	82.4	6.3	46.7
	5	116	88.4	32.0	11.0	85.9	3.1	54.0
	6	108	87.5	31.9	10.6	86.9	2.5	53.0
	Mean	116 ± 14	87.1 ± 1.4	31.5 ± 1.0	10.9 ± 0.7	85.1 ± 1.6	4.0 ± 1.6	48.6 ± 4.7
Y 75 mm	1	99	75.0	34.1	11.2	85.9	2.9	55.6
	2	97	85.8	34.6	11.6	85.1	3.3	53.7
	3	103	86.0	34.9	12.2	84.2	3.6	38.2
	4	98	87.3	34.7	11.3	85.5	3.2	40.8
	5	120	84.4	35.0	13.9	83.1	3.0	47.6
	6	110	80.9	33.6	12.3	85.6	2.1	50.3
	Mean	105 ± 9	83.2 ± 4.6	34.5 ± 0.5	12.1 ± 1.0	84.9 ± 1.1	3.0 ± 0.5	47.7 ± 7.0

In Figure 6a,b, SEM micrographs of a pearlite island in GJS 400 from Y 25 mm sample are reported. The clear lamellar pattern, i.e., parallel lamellae at an almost uniform distance, that can be seen in Figure 6b is not frequent, since pearlite often shows a complex configuration, in which the lamellar structure is irregular. Therefore, the characteristic widths of ferritic channels in the pearlitic islands could not be measured and can only be estimated to span between 100 and 300 nm, independently of cooling rates.

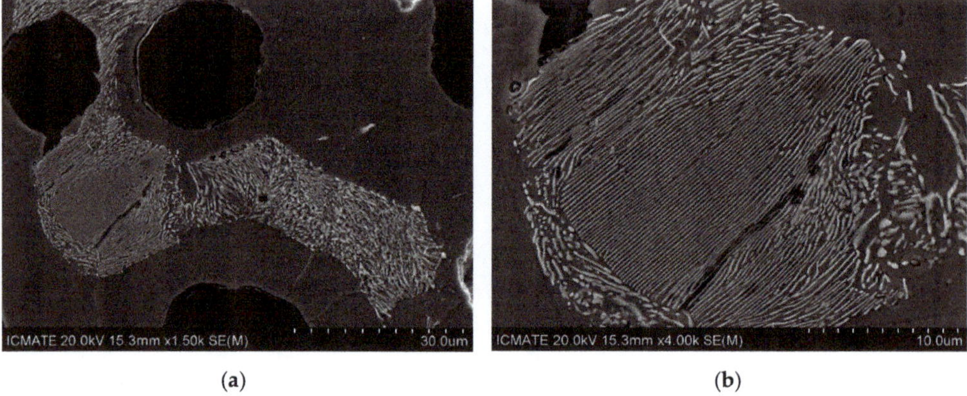

(a) (b)

Figure 6. SEM micrographs (SEI) of a typical pearlitic island in GJS 400 (Y 25 mm) with lamellar regions with ferritic channels of nanometric widths and irregular pearlite at different magnifications: (**a**) 1500 X; (**b**) 4000 X.

3.3. EDS Analyses

The local chemical composition of GJS 400 specimens from the four different samples was investigated through EDS. In particular, the concentration gradient of Si and Mn between couples of graphitic nodules was considered. Results are significantly different whether or not pearlite is present. Figure 7 shows a typical example of Si and Mn content in the region between two nodules separated by a pearlitic island (Y 75 mm sample). The Mn enrichment (positive segregation) and Si depletion (negative segregation) throughout pearlite is a common feature shown by every specimen, independently of the mold geometry. When there is no pearlite, neither Si nor Mn shows composition gradient (Figure 8).

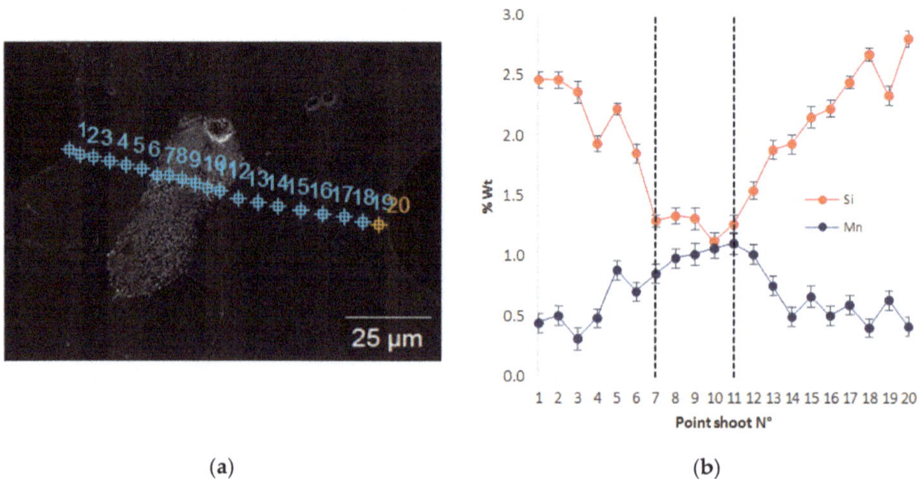

(a) (b)

Figure 7. Energy Dispersive X-ray Spectroscopy (EDS) investigation through a pearlitic island in GJS 400 (Y 75 mm sample): (**a**) EDS point shots positions; (**b**) gradients of Si and Mn compositions (wt.%) versus EDS point positions.

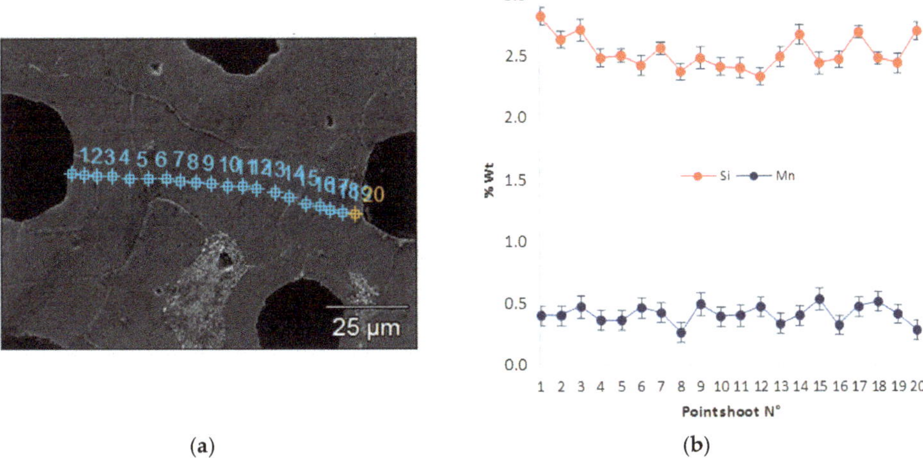

(a) (b)

Figure 8. EDS investigation through ferrite in GJS 400 (Y 75 mm sample): (**a**) EDS point shots positions; (**b**) gradients of Si and Mn compositions (wt.%) versus EDS point positions.

It has to be pointed out that the EDS probe overestimated the Mn content, which is about 0.1% (Table 1). This is thought to be an issue of EDS analysis itself, since it is difficult to determine the quantity of trace elements (concentration lower than about 1% wt). Mn content is indeed low and this could affect the absolute values given by the EDS measurements. Its gradient, though, can be considered significant.

4. Discussion

The GJS 400 microstructures are consistent with the simulated solidification rates (Figure 4), so that microstructural features result finer when cooling rates are higher (Table 3), in agreement with what reported in literature [10,31]. Nodule count measurement as a function of cooling rate at T_s (Figure 9) is consistent with the relationship found by Górny et al. in ductile iron with no Cu addition [10]. The presence of Cu in the alloys investigated in this work could account for the increase of nodule count at the same cooling rate.

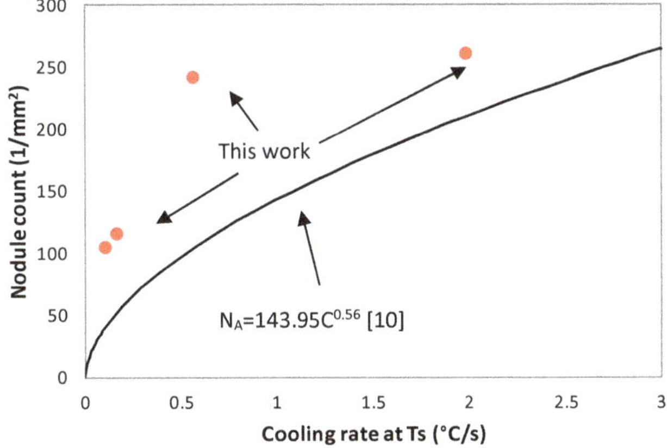

Figure 9. Nodule count (N_A) as a function of cooling rate (C) at the eutectic temperature T_s (red dots). The black line represents the relationship between cooling rate and nodule count in [10].

Higher cooling rates around eutectic temperature also lead to higher undercooling, which can be in turn fairly related to nodule count and nodule mean diameter (Figure 10).

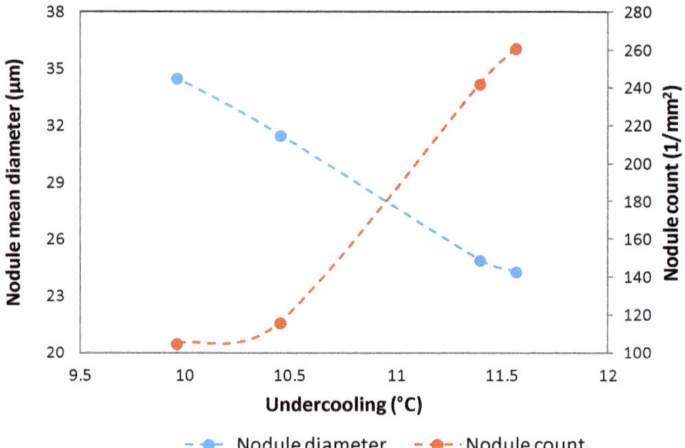

Figure 10. Nodule mean diameter and nodule count as functions of undercooling (difference between the eutectic temperature according to Equation (1) and the minimum temperature at the beginning of solidification).

Volume fraction of graphite, nodule count, and the mean nodule diameter, listed in Table 3, can be used to calculate the mean distance λ between graphite nodules through the Fullman's equation [32]:

$$\lambda = \frac{1 - V_g}{d N_A},$$

(3)

where V_g is the volume fraction of graphite, N_A is the nodule count, and d is the mean diameter of the nodules.

The mean values for the four molds calculated through Equation (3) are reported in Table 4.

Table 4. Mean distance between graphitic nodules according to Equation (3).

Sample	Lynchburg	Y 25 mm	Y 50 mm	Y 75 mm
λ (μm)	136.2	144.4	243.8	242.7

The mean value for the Y 50 specimens is slightly higher than the one for the Y 75, despite the higher cooling rate, mainly because of the higher graphite volume fraction (Table 3).

The graphite content (Table 3, V_g in Equation (3)) is consistent with the Wojnar estimation [33] based on the carbon content of the alloy:

$$V_g = \frac{7.8\%C}{222 + 5.6\%C}.$$

(4)

Being %C the weight content of the alloy (3.63%), Equation (4) predicts a graphite volume fraction of 11.7%.

As already found in literature [20], ferritic grain size decreases when solidification rate increases (Figure 11).

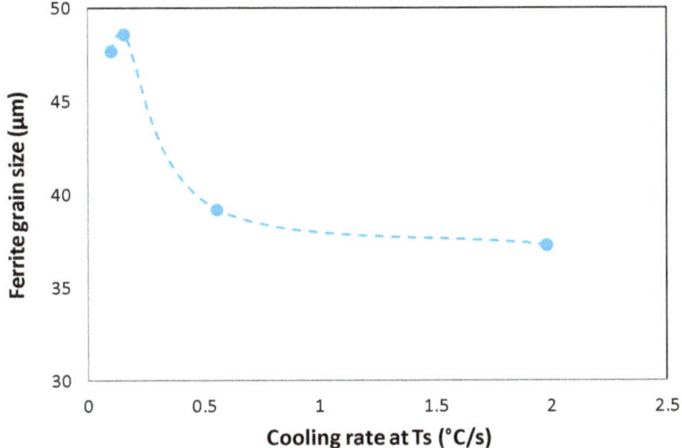

Figure 11. Ferrite grain size as a function of cooling rate at the eutectic temperature T_s.

While no apparent composition gradients can be seen in ferritic grains, pearlitic islands show positive Mn segregation and negative Si segregation (Figures 7 and 8), in agreement with literature [34–36]. These gradients are related to what happens during eutectic solidification. Mn is continuously rejected from the solidification front to the melt metal, making Mn content increase in the last to freeze zone, namely, the grain boundaries between nodules. Mn, as well as other carbide forming elements like Cr and V at the left side of Fe in the periodic table, promotes pearlite, which explains why it is found in pearlitic islands.

On the other hand, Si, which promotes graphite formation in the melt metal, tends to remain in the first to freeze zone, around the graphite nodules, promoting ferrite.

After solidification, solid state transformations take place. In particular, ferrite nucleates and grows in austenitic grains, which transforms into ferrite and graphite. If cooling is fast enough, thus allowing for larger undercooling, the eutectoid transformation occurs and pearlite forms [2,14].

Table 2 and Figure 4b show that cooling rates at the eutectoid temperature were low for all the four samples, and this is consistent with the very low pearlite volume fractions found. In the Lynchburg mold pearlite it is barely detectable, even if the cooling rate at the eutectoid transformation was higher than in Y 50 and Y 75 samples. This suggests that a major role was played by cooling rate at solidification which, at the eutectic temperature, was much higher in the Lynchburg mold. This may have reduced Mn segregations, thus lowering the pearlite content. So, pearlite may be the product of segregations during solidification rather than the result of different cooling rates through the intercritical interval Ar1-Ar3.

5. Conclusions

Different microstructures of GJS 400 were obtained through different geometries leading to different cooling rates, which were calculated through simulation of the actual gravity casting system. The microstructures were characterized in details, quantifying nodule count, nodularity, average diameter of the graphitic nodules, and volumetric fractions of graphite and pearlite compliant with the minimum requirements of statistics of the standard ASTM E2567-16a [28], and the average ferritic grain size complying with the standard ASTM E112-13 [29]. These features result finer as the solidification rate increases.

Positive segregation (enrichment) of Mn and negative segregation (depletion) of Si was observed in the pearlitic islands.

The cooling rates around the eutectoid temperature were very similar and very low, which prevented pearlite formation. Data suggest that the occurrence of pearlite is related to segregations during solidification, rather than to cooling rates at the eutectoid temperature.

This microstructural characterization provides the basis for the description and modeling of the tensile properties of GJS 400 alloy, the subject of a second part of this investigation [25].

Author Contributions: Conceptualization, G.A.; methodology, G.A.; software, S.M.; investigation, G.A., D.R., S.M.; resources, G.A., F.Z.; data analysis, G.A., D.R., S.M.; writing—original draft preparation, G.A.; writing—review and editing, D.R., M.G.; supervision, G.A.

Funding: This research received no external funding.

Acknowledgments: The authors would like to thank Davide Della Torre, Tullio Ranucci and Marcello Taloni for their experimental support.

Conflicts of Interest: The authors declare no conflict of interest.

References

1. Labrecque, C.; Gagné, M. Review Ductile Iron: Fifty Years of Continuous Development. *Can. Metall. Quart.* **1998**, *37*, 343–378. [CrossRef]
2. Tiedje, N.S. Solidification, Processing and Properties of Ductile Cast Iron. *Mater. Sci. Tech. Ser.* **2010**, *26*, 505–514. [CrossRef]
3. Fraś, E.; López, H. Eutectic Cells and Nodule Count—An Index of Molten Iron Quality. *Int. J. Metalcast.* **2010**, *4*, 35–61. [CrossRef]
4. Showman, R.E.; Aufderheide, R. A process for thin-wall sand castings. *AFS Trans.* **2003**, *111*, 567–578.
5. Juretzko, F.R.; Dix, L.P.; Ruxanda, R.; Stefanescu, D.M. Precondition of Ductile Iron Melts for Light Weight Casting: Effect on Mechanical Properties and Microstructure. *AFS Trans.* **2004**, *112*, 773–785.
6. Fraś, E.; López, H.; Podrzucki, C. The Influence of Oxygen on the Inoculation Process of Cast Iron. *Int. J. Cast Metal. Res.* **2000**, *13*, 107–121. [CrossRef]
7. Popescu, M.; Zavadil, R.; Thomson, J.; Sahoo, M. Studies to Improve Nucleation Potential of Ductile Iron When Using Carbidic Ductile Iron Returns. *AFS Trans.* **2007**, *115*, 591–608.
8. Fraś, E.; Górny, M. Fading of inoculation effects in ductile iron. *Arch. Foundry Eng.* **2008**, *8*, 83–88.
9. Fraś, E.; Górny, M.; López, H.F. Eutectic cell and Nodule Count in Cast Iron. Part I. Theoretical Background. *ISIJ Int.* **2007**, *47*, 259–268. [CrossRef]
10. Górny, M.; Tyrała, E. Effect of Cooling Rate on Microstructure and Mechanical Properties of Thin-Walled Ductile Iron Castings. *J. Mater. Eng. Perform.* **2013**, *22*, 300–305. [CrossRef]
11. White, D. Production of Ductile Iron Castings. In *ASM Handbook, Volume 1A: Cast Iron Science and Technology*; ASM International: Materials Park, OH, USA, 2017; pp. 603–611. [CrossRef]
12. Skaland, T.; Grong, O. Nodule Distribution in Ductile Cast Iron. *AFS Trans.* **1991**, *99*, 153–157.
13. Venugopalan, D. Prediction of Matrix Microstructure in Ductile Iron. *AFS Trans.* **1990**, *98*, 465–469.
14. Gerval, V.; Lacaze, J. Critical Temperature Range in Spheroidal Graphite Cast Irons. *ISIJ Int.* **2000**, *40*, 386–392. [CrossRef]
15. Górny, M.; Angella, G.; Tyrała, E.; Kawalec, M.; Paź, S.; Kmita, A. Role of Austenitization Temperature on Structure Homogeneity and Transformation Kinetics in Austempered Ductile Iron. *Met. Mater. Int.* **2019**, *25*, 956–965. [CrossRef]
16. Donnini, R.; Fabrizi, A.; Bonollo, F.; Zanardi, F.; Angella, G. Assessment of the microstructure evolution of an austempered ductile iron during austempering process through strain hardening analysis. *Met. Mater. Int.* **2017**, *23*, 855–864. [CrossRef]
17. Alhussein, A.; Risbet, M.; Bastien, A.; Chobaut, J.P.; Balloy, D.; Favergeon, J. Influence of silicon and addition elements on the mechanical behaviour of ferritic ductile cast iron. *Mater. Sci. Eng. A-Struct.* **2014**, *605*, 222–228. [CrossRef]
18. Lacaze, J.; Boudot, A.; Gerval, A.; Oquab, D.; Santos, H. The Role of Manganese and Copper in Eutectoid Transformation of Spheroidal Graphite Cast Iron. *Metall. Mater. Trans. A* **1997**, *28*, 2015–2025. [CrossRef]
19. Angella, G.; Zanardi, F.; Donnini, R. On the significance to use dislocation-density-related constitutive equations to correlate strain hardening with microstructure of metallic alloys: The case of conventional and austempered ductile irons. *J. Alloys Compd.* **2016**, *669*, 262–271. [CrossRef]
20. Hütter, G.; Zybell, L.; Kuna, M. Micromechanisms of fracture in nodular cast iron: From experimental findings 466 towards modeling strategies—A review. *Eng. Fract. Mech.* **2015**, *144*, 118–141. [CrossRef]

21. Goodrich, G.M. Cast iron microstructure anomalies and their causes. *AFS Trans.* **1997**, *105*, 669–683.
22. Iwabuchi, Y.; Narita, H.; Tsumura, O. Toughness and Ductility of heavy-walled ferritic spheroidal-graphite iron castings. *Res. Rep. Kushiro Natl. Coll.* **2003**, *37*, 1–9.
23. Nilsson, K.F.; Blagoeva, D.; Moretto, P. An experimental and numerical analysis to correlate variation in 472 ductility to defects and microstructure in ductile cast iron components. *Eng. Fract. Mech.* **2006**, *73*, 1133–1157. [CrossRef]
24. Nilsson, K.F.; Vokal, V. Analysis of ductile cast iron tensile tests to relate ductility variation to casting defects 475 and material microstructure. *Mater. Sci. Eng. A-Struct.* **2009**, *502*, 54–63. [CrossRef]
25. Angella, G.; Ripamonti, D.; Górny, M.; Masaggia, S.; Zanardi, F. The role of the microstructure on the tensile plastic behaviour of the ductile iron GJS 400 produced through different cooling rates. *Metals* **2019**, *9*, 1019. [CrossRef]
26. *EN 1563, Founding – Spheroidal Cast Irons*; European Committee for standardization (CEN): Brussels, Belgium, 2018.
27. ImageJ 1.51i. Available online: https://imagej.net (accessed on 30 March 2018).
28. *ASTM E2567-16a, Standard Test Method for Determining Nodularity and Nodule Count in Ductile Iron Using Image Analysis*; ASTM International: West Conshohocken, PA, USA, 2016. [CrossRef]
29. *ASTM E112-13, Standard Test Methods for Determining Average Grain Size*; ASTM International: West Conshohocken, PA, USA, 2013. [CrossRef]
30. Neuman, F. The influence of additional elements on the physic-chemical behaviour of carbon in saturated molten iron. In *Recent Research on Cast Iron*; Gordon and Breach: New York, NY, USA, 1998; pp. 659–705.
31. Rivera, G.; Boeri, R.; Sikora, J. Influence of the inoculation process, the chemical composition and the cooling rate, on the solidification macro and microstructure of ductile iron. *Int. J. Cast Metal. Res.* **2003**, *16*, 23–28. [CrossRef]
32. Fullman, R.L. Measurement of Particle Sizes in Opaque Bodies. *Trans. AIME* **1953**, *197*, 447–452. [CrossRef]
33. Wojnar, L.; Kurzydłowski, K.J.; Szala, J. Quantitative Image Analysis [for metallography]. In *Metallography and Microstructures*; ASM International: Materials Park, OH, USA, 2004; pp. 403–427.
34. Guo, X.; Stefanescu, D.M. Partitioning of alloying elements during the eutectoid transformation of ductile iron. *Int. J. Cast Metal. Res.* **1999**, *11*, 437–441. [CrossRef]
35. Boeri, R.; Weinberg, F. Microsegregation of Alloying Elements in Cast Iron. *Int. J. Cast Metal. Res.* **1993**, *6*, 153–158. [CrossRef]
36. Chiniforush, E.A.; Iranipour, N.; Yazdani, S. Effect of nodule count and austempering heat treatment on segregation behavior of alloying elements in ductile cast iron. *China Foundry* **2016**, *13*, 217–222. [CrossRef]

Article

The Role of Microstructure on the Tensile Plastic Behaviour of Ductile Iron GJS 400 Produced through Different Cooling Rates—Part II: Tensile Modelling

Giuliano Angella [1],*, Riccardo Donnini [1], Dario Ripamonti [1], Marcin Górny [2] and Franco Zanardi [3]

[1] National Research Council of Italy (CNR)—Institute of Condensed Matter Chemistry and Technologies for Energy (ICMATE), via R. Cozzi 53, 20125 Milan, Italy; riccardo.donnini@cnr.it (R.D.); dario.ripamonti@cnr.it (D.R.)

[2] Faculty of Foundry Engineering, Department of Cast Alloys and Composites Engineering, AGH University of Science and Technology, 30-059 Krakow, Poland; mgorny@agh.edu.pl

[3] Zanardi Fonderie S.p.A., via Nazionale 3, 37046 Minerbe (VR), Italy; franco.zanardi@icloud.com

* Correspondence: giuliano.angella@cnr.it; Tel.: +39-02-66173-327

Received: 30 August 2019; Accepted: 17 September 2019; Published: 19 September 2019

Abstract: Tensile testing on ductile iron GJS 400 with different microstructures produced through four different cooling rates was performed in order to investigate the relevance of the microstructure's parameters on its plastic behaviour. Tensile flow curve modelling was carried out with the Follansbee and Estrin-Kocks-Mecking approach that allowed for an explicit correlation between plastic behaviour and some microstructure parameters. In the model, the ferritic grain size and volume fraction of pearlite and ferrite gathered in the first part of this investigation were used as inputs, while other parameters, like nodule count and interlamellar spacing in pearlite, were neglected. The model matched very well with the experimental flow curves at high strains, while some mismatch was found only at small strains, which was ascribed to the decohesion between the graphite nodules and the ferritic matrix that occurred just after yielding. It can be concluded that the plastic behaviour of GJS 400 depends mainly on the ferritic grain size and pearlitic volume fraction, and other microstructure parameters can be neglected, primarily because of their high nodularity and few defects.

Keywords: ductile cast irons; tensile tests; microstructure; plasticity modelling

---

1. Introduction

Ductile Irons (DIs) are cast irons containing graphite of a spherical shape, which gives them an excellent compromise between tensile strength and ductility, toughness, and fatigue resistance [1–6]. Thanks to their fine microstructure control, DIs can satisfy a variety of different design requirements for heavy duty components, such as in hydraulic and oleo dynamic applications, as well as bearing adapters. However, in components with complex geometry, the solidification conditions can be quite different, producing different microstructures and, as a consequence, different plastic behaviours. Thus, the relationship between solidification conditions, microstructure, and plastic behaviour have to be known in order to optimise the design of the component's geometry [7–11]. The strain hardening behaviour and strength of DIs are strongly affected by their microstructures, since the yield stress increases with decreasing ferritic grain size, while in ferritic–pearlitic DIs, the yield stress increases with an increase in pearlite content, which causes a reduction of the strain to fracture [12]. With low nodularity, the DIs become more brittle, with decreasing strains at failure [13–16]. Empirical relations for the correlations between the mechanical properties and microstructure in DIs were reported in [15,16].

The present paper is focused on investigating the relevance of the microstructure parameters on the plastic behaviour of GJS 400 with different microstructures produced through four different cooling rates. The details of the GJS 400 microstructure investigations are reported elsewhere [17]. The tensile flow curve analysis was carried out with the Follansbee and Estrin-Kocks-Mecking approach, which allows an explicit correlation between the plastic behaviour and the microstructure parameters.

2. Materials and Methods

2.1. Material

The chemical compositions of the GJS 400 produced in Zanardi Fonderie S.p.A. (Minerbe-VR, Italy), with four different moulds, is reported in Table 1.

Table 1. Chemical composition of GJS 400 in wt %.

C	Si	Mg	Mn	Cu	Ni	Cr	P	S	Fe
3.63	2.45	0.046	0.129	0.133	0.0168	0.023	0.038	0.0043	Bal.

The GJS 400 was produced using four different cooling rates with a cylindrical Lynchburg (25 mm diameter), and 3 different Y moulds with increasing thicknesses of 25, 50, and 75 mm, thereby complying with the standard ASTM A 536-84. Details of the melt pouring conditions and cooling rates have been reported in [17], where the microstructures of the 24 GJS-400 samples (6 samples from each mould) were investigated. The nodule count, nodularity, average diameter of the graphite nodules, volumetric fractions of graphite, and pearlite were measured by digital image analysis complying with ASTM E2567-16a, and the average ferritic grain size was also found (complying with ASTM E112-13). An example of a typical GJS 400 microstructure produced through a Y 75 mm mould is reported in Figure 1 after etching with Nital 2%, where spheroidal graphite (black) in the ferritic matrix with visible grain boundaries and pearlitic islands (light grey) are visible. The average parameters of the microstructures of the GJS 400 produced with the four moulds are reported in Table 2, while the microstructure parameters from each sample are reported elsewhere [17]. In the Lynchburg samples, there was no significant pearlite, while in the Y moulds where the cooling rates were slower, the pearlite volume fractions spanned from about 3% to 4%. However, the pearlite formation was rationalized in term of the positive micro-segregations of Mn and the negative micro-segregations of Si that were produced during solidification [17–19]. The ferritic grain size range was from 37.3 to 48.6 μm, which is consistent with the solidification rates.

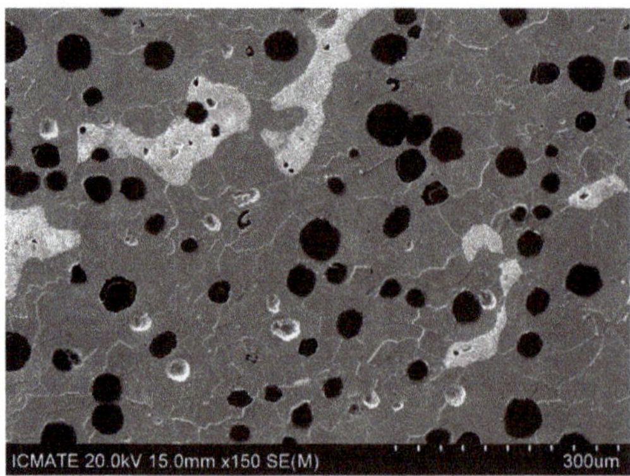

Figure 1. Scanning electron microscopy (SEM) micrograph with secondary electron imaging of GJS 400 produced with the Y 75 mm mould after etching with Nital 2%: nodular graphite (black) in the ferritic matrix with bright grain boundaries and pearlitic islands (light grey).

Table 2. Average parameters of the GJS 400 microstructures produced with four different moulds. The errors associated with the measurements are the standard deviations.

Mould	Nodule Count (mm^{-2})	Nodule Size (µm)	Nodularity (%)	Pearlite Volume Fraction (%)	Ferrite Grain Size (µm)
Lynchburg 25 mm	261 ± 15	24.3 ± 0.6	89.8 ± 3.0	-	37.3 ± 3.2
Y 25 mm	242 ± 11	24.9 ± 0.5	91.2 ± 1.6	3.8 ± 0.4	39.2 ± 2.3
Y 50 mm	116 ± 14	31.5 ± 1.0	87.1 ± 1.4	4.0 ± 1.6	48.6 ± 4.7
Y 75 mm	105 ± 9	34.5 ± 0.5	83.2 ± 4.6	3.0 ± 0.5	47.5 ± 7.2

2.2. Tensile Tests and Microstructure Plasticity Model

Tensile tests on the considered 24 samples of GJS 400 were carried out on round specimens using a gauge with an initial diameter of d_0 = 12.5 mm and a length of l_0 = 50 mm, complying with the standard ASTM E8-8M with a strain rate of 10^{-4} s^{-1}. The true stress–true plastic strains (σ vs. ε_p) were used, where $\sigma = S \cdot (1 + e)$ and $\varepsilon_p = \varepsilon - \sigma/E = \ln(1 + e) - \sigma/E$ (with S and e as the engineering stress and elongation, respectively) and E is the experimental Young modulus.

The model used to describe the tensile plastic flow curves of the GJS 400 with a detailed correlation with the microstructure's characteristic lengths was based on the Follansbee approach [20,21], according to which the flow stress σ develops as

$$\sigma = \sigma_o + \sigma_G(\varepsilon_P) = \sigma_o + M\alpha_o Gb\rho^{1/2}, \tag{1}$$

where σ_o is the initial stress because of the solid solution or precipitation strengthening, and σ_G is the component of stress depending on the increase of the dislocation density ρ because of strain ε_P. M is the Taylor factor (3.01 in BCC materials), α_o the dislocation-dislocation interaction strength (0.5) [22], G the elastic shear modulus for ferrite (64 GPa), and b the Burgers vector length of ferrite (0.248 nm). At strain ε_P = 0, σ_G was assumed to be nil because of the negligible initial dislocation density of GJS 400 in the cast conditions. The total dislocation-density ρ increases because of straining, according to the mechanistic evolution equation by Kocks-Mecking-Estrin [22–28]:

$$\frac{d\rho}{d\varepsilon_P} = M \cdot \left[\left(\frac{1}{b\Lambda} + \frac{1}{bD} + \frac{1}{b\lambda} \right) - D_o \cdot \rho \right], \tag{2}$$

where D_o is the dynamic recovery term that describes the softening of materials during straining because of dislocation annihilation and low energy dislocation structure formation. Λ, D, and λ are the microstructure characteristic lengths; Λ is the dislocations mean free path related to the dislocation cells in ferrite with $\Lambda = \beta/\rho^{1/2}$ and a β constant [22], D is the ferritic grain size or pearlitic island size, and λ is the interlamellar spacing in pearlite.

Substituting $\sigma_G = M\alpha_o Gb\rho^{1/2}$ and $\Lambda = \beta/\rho^{1/2}$, and considering that in ferrite the grain boundaries and dislocation cells are the obstacles to dislocation motion (for ferrite Equation (2)), results in

$$\frac{d\sigma_G}{d\varepsilon_P} = \left(\frac{K_o}{\beta} + \frac{K_1}{D \cdot \sigma_G} \right) - \frac{\sigma_G}{\varepsilon_{c,F}}, \tag{3}$$

where K_o (= 1.538 × 10^5 MPa) and K_1 (= 7.565 × 10^6 MPa$^2 \cdot$µm) are constants depending on the BCC ferritic crystal, while β and $1/\varepsilon_{c,F}$ are outputs from fitting. The detailed calculations to obtain Equation (3) from Equations (1) and (2) are reported in Appendix A.

In pearlite, the interlamellar spacing λ is nanometric, which is by far smaller than the Λ, the ferritic grain size D_{Ferrite}, and the pearlitic colony size, D_{Pearlite}. Thus, pearlite Equation (2) results in

$$\frac{d\sigma_G}{d\varepsilon_P} = \frac{K_1}{\lambda \cdot \sigma_G} - \frac{\sigma_G}{\varepsilon_{c,P}} \tag{4}$$

where $1/\varepsilon_{c,P}$ is the output from the fitting, if λ is known. The detailed calculations to obtain Equation (4) from Equations (1) and (2) are reported in Appendix A. The equation σ_G vs. ε_P, resulting from integrating Equation (4), is an exponential decay equation with a saturation stress $\sigma_{S,P}$ that is the maximum stress achieved asymptotically at the condition $d\sigma_G/d\varepsilon_P = 0$, while $\varepsilon_{c,P}$ is the critical strain that defines the rate at which $\sigma_{S,P}$ is achieved. However, if an average characteristic λ cannot be measured (like in the present investigation of GJS 400 because of the complexity of pearlitic microstructures [17]), Equation (4) can be fitted to the experimental data considering the quantity (K_1/λ) as a further output from the fitting. Then, from (K_1/λ) and $\varepsilon_{c,P}$, the saturation stress $\sigma_{S,P} = [(K_1/\lambda) \cdot \varepsilon_{c,P}]$ can be found to test the physical meaning of the fitting results.

GJS 400 are cast irons with different volume fractions of ferrite and pearlite, resulting from the solidification rates, as reported in Table 2. Thus, the total tensile flow stress with strain $\sigma(\varepsilon_P)$ in GJS 400 produces a mixture rule:

$$\sigma(\varepsilon_P) = (1 - X_{Pearlite}) \cdot \sigma_{Ferrite}(\varepsilon_P) + X_{Pearlite} \cdot \sigma_{Pearlite}(\varepsilon_P) \tag{5}$$

where $X_{Pearlite}$ is the pearlite volume fraction, and $(1 - X_{Pearlite})$ is the ferrite volume fraction. The rule of mixture that has been usually used for all two phase materials [29] has also been successfully used in Dual Phase (DP) steels [30–32] whose microstructures consist of soft ferrite and hard martensite. Equation (5) was used successfully in DP steels for hardness, Yield Stress (YS), and Ultimate Tensile Stress (R_m). In terms of mechanical constituents, DP's microstructure has similarities with the investigated GJS 400, consisting of soft ferrite and hard pearlite (and graphite nodules), so Equation (5) was used for the present investigation.

The fundamental assumption of this approach is that the graphite should not affect the tensile plastic behaviour of GJS 400, so graphite parameters like nodule count, nodule size, internodular spacing, and nodularity were expected to not be needed in the first approximation to describe the tensile plastic behaviour. This assumption had to be validated.

3. Results

3.1. Model Calibration

The model was calibrated firstly by fitting the tensile flow curves of the GJS 400 from Lynchburg mould samples with Equation (3), where only ferrite was found (see Table 2). In this way, the fitting parameters concerning ferrite were found as outputs from the fitting. The second calibration step was used to fit the tensile data of GJS 400 from Y 25 mm with Equation (5), where the pearlite was also present, in order to work out the flow curve of the pearlite. Thus, after the latter step, the microstructure plasticity model was calibrated, and then the tensile flow curves of the GJS 400 from moulds Y 50 mm and 75 mm could be modelled using only their microstructure parameters reported in Table 2, and then comparing them to the experimental flow curves.

By analysing the six tensile flow curves of the GJS 400 produced with the Lynchburg mould in order to have the best strain hardening fittings at high stresses, an average initial stress $\sigma_0 = 243.1 \pm 6.2$ MPa (see Equation (1)) was found. Indeed, the plastic flow curves of GJS 400 from different moulds did not change during the early stages of deformations but was significant at high strains, which could be rationalized by the findings that ferrite was the dominant softer constituent that deformed first at yielding, while the smaller volume fractions (<4%) of harder pearlite contributed significantly later at higher strains. Thus, $\sigma_0 = 243.1$ MPa was used to model all the flow curves of GJS 400 from the other moulds. The strain hardening data of the tensile flow curves of GJS 400 from the Lynchburg mould with an average ferrite grain size of 37.3 ± 3.2 µm were fitted, yielding the following average values for the equation parameters: $1/\varepsilon_{c,F} = 6.36 \pm 0.25$ and $\beta = 119.1 \pm 8.7$ MPa (see Equation (3)). These values are consistent with the literature, where β has been reported to be between 100 and 200 [22], proving the physical meaning and, in turn, the validity of the model.

In the GJS 400 samples produced in the present investigation, the pearlite was irregular and its shape was rarely lamellar and depended on grain orientation [17], so it was not possible to measure any characteristic average interlamellar spacing λ to input into Equations (4) and (5). Thus, in order to find the pearlitic flow curve $\sigma_{\text{Pearlite}}(\varepsilon_P)$ vs. ε_P, Equation (5) was fitted to the GJS 400 Y 25 mm tensile data considering the quantity (K_1/λ) and the parameter $1/\varepsilon_{c,P}$ as outputs. In the GJS 400 produced with Y 25 mm, the average volume fraction of the pearlite was 3.8% ± 0.4%, and the average grain size was 39.2 ± 2.3 µm (see Table 2). The fitting resulted in a pearlite flow curve with an average saturation stress of $\sigma_{S,P}$ = 1094.4 ± 106.0 MPa and an average critical strain parameter of $1/\varepsilon_{c,P}$ = 22.1 ± 3.3. In Figure 2a,b, the fitting results are reported for a typical flow curve of GJS 400 from the Y 25 mm mould sample, detailing the contributions from the ferrite and pearlite. The fit was excellent at high stresses, while at low strains there was some mismatch. Though $\sigma_{S,P}$ was consistent with the results reported in the literature for Isothermed Ductile Irons 1000 with a pearlite volume fraction higher than 80% [27], the $1/\varepsilon_{c,P}$ for pearlite was quite low, considering that it should have been just slightly lower than 40. In other words, the pearlite contribution to the flow curve in Figure 2b should have increased faster while keeping the same saturation stress, $\sigma_{\text{Pearlite},V}$. The reasons for this result are not evident and need further investigation.

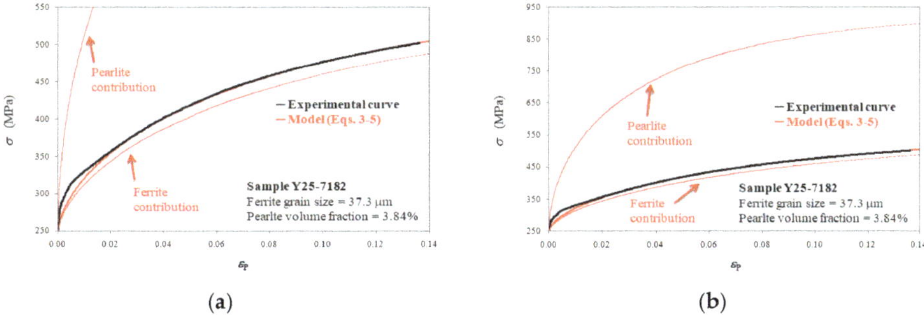

Figure 2. (**a**) Example of fitting Equation (5) with a typical tensile flow curve of GJS 400 from the Y 25 mm mould sample; (**b**) the same fitting in (**a**) at a different scale to highlight the contributions from the pearlite.

3.2. Model Prediction

After calibration, Equations (3)–(5) were used to predict the tensile flow curves of the GJS 400 produced from the moulds Y 50 mm and Y 75 mm by using only the microstructure parameters for the ferritic grain sizes and pearlite volume fractions, reported in Table 2. Examples of typical model curves are reported in Figure 3a–d for GJS 400 produced from the Y 50 mm and Y 75 mm moulds, respectively, where only the ferritic component contributions are reported. The ferrite grain size and pearlite volume fraction of the individual samples are reported on the plots. Indeed, there is a significant mismatch, even though the model can qualitatively describe the experimental flow curves.

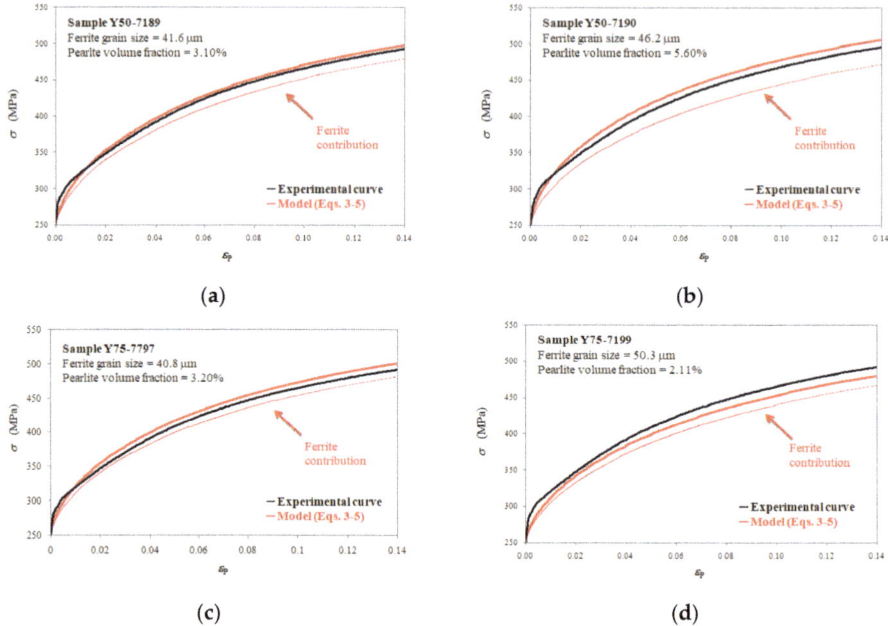

Figure 3. Examples of modelling with Equations (3)–(5) the tensile flow curve of the GJS 400 from different moulds: (**a**) and (**b**) from Y 50 mm; (**c**) and (**d**) from Y 75 mm.

4. Discussion

4.1. Considerations of the Minimum Requirements of Data Statistics Complying with the Standards ASTM E2567-16a and ASTM E112-13

The model flow curves reported in Figure 3 can qualitatively predict the experimental tensile behaviour of GJS 400 produced through different cooling rates (Y 50 mm and Y 75 mm moulds), suggesting that, though the microstructure plasticity model was qualitatively correct, it could be improved. In Figure 3, the individual microstructure parameters of the GJS 400 samples from the Y 50 mm and Y 75 mm moulds were used as inputs in Equations (3)–(5). However, if the average values of the ferrite grain sizes (48.6 μm in the Y 50 mm mould, and 47.7 μm in the Y 75 mm mould) and pearlite volume fractions (4.0% in the Y 50 mm mould, and 3.0% in the Y 75 mm mould) from Table 2 are used, the matches between the model flow curves and the experimental data improve significantly, albeit at small strains. In Figure 4, the same experimental flow curves of Figure 3 were compared to the model flow curves where the average microstructure parameters in Table 2 were used.

Indeed, the local microstructure parameters in Figure 3 changed significantly from the sample Y50-7189 to Y50-7190 (the Y 50 mm mould). The ferritic grain size changed from 41.6 μm to 46.2 μm, and the pearlite volume fraction changes from 3.1% to 5.6%. The same wide change was found for the samples from the Y 75 mm mould. The ferritic grain size changed from 40.8 μm to 50.3 μm, and the pearlite volume fraction changed from 3.2% to 2.1%. These results suggested that the average microstructure parameters became more adherent to the actual microstructures than the single microstructure parameters gathered from each sample.

The rationalization of this finding is that the reason for the mismatch between the experimental curves and the model in Figure 3 was the inaccuracy of the microstructure parameters, although they were carried out according to the minimum requirements of the data statistics, complying with the standard ASTM E112-13 for the grain size measurements reported in [17]. Indeed, the average microstructure parameters calculated for the 6 samples of each mould represented a statistical increase of six times with respect to the minimum required statistics. Thus, the GJS 400 microstructures were significantly more homogeneous than those found through the single sample measurements and were better described by the average microstructure parameters in Table 2. On the other hand, the mechanical tensile tests were indirect characterizations of the microstructures complying with the minimum data statistics that proved to be more reliable than the direct characterizations required by the ASTM E2567-16a and ASTM E112-13 standards. This result means that if wider statistics were gathered beyond the minimum requirements of the standards' statistics, the accuracy of the microstructure parameters from each sample could be improved.

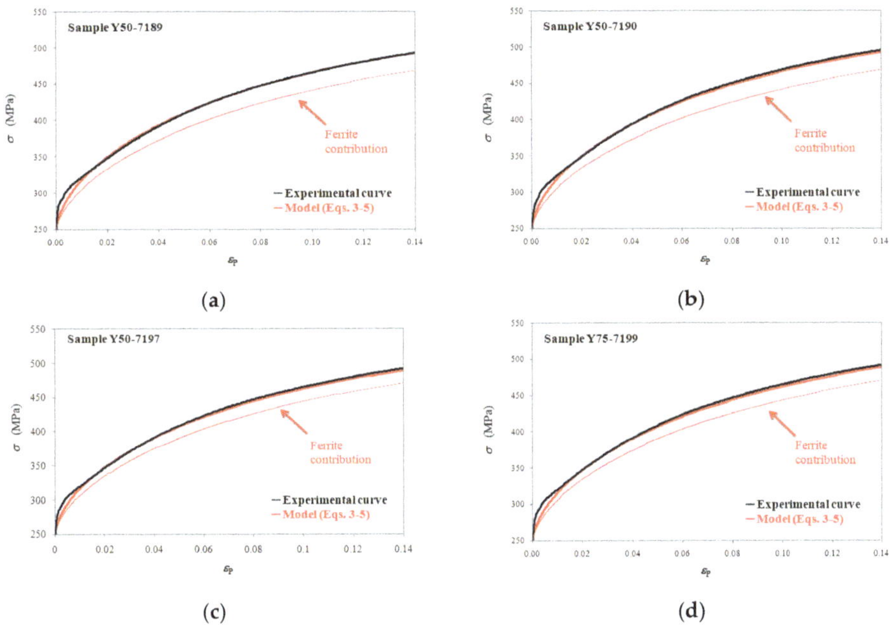

Figure 4. Modelling the tensile flow curves of the GJS 400 reported in Figure 3 using the average microstructure values in Table 2: (**a**) and (**b**) Y 50 mm with the average ferrite grains size = 48.6 μm, and average pearlite volume fraction = 4.0%; (**c**) and (**d**) Y 75 mm with average ferrite grains size = 47.7 μm, and average pearlite volume fraction = 3.0%.

4.2. Microstructure Parameters Relevant to Describing the Plastic Behaviour of GJS 400

As seen in Figure 4, using the average microstructure parameters (see Table 2) as inputs in the microstructure plasticity model the result excellently described the experimental flow curves of the GJS 400 produced with different cooling rates, even if minor mismatch was present at small strains. Thus, the microstructure plasticity model indicated that the ferritic grains size and pearlite volume fractions were the only microstructural parameters needed to describe the plastic behaviour of GJS 400 produced in the range of the cooling rates tested with the Lynchburg and Y 25–75 mm moulds. The model flow curves in the engineering stress–strain coordinates (up to an ultimate tensile stress R_m) built with the average microstructure parameters for the four different moulds in Table 2 are reported in Figure 5, while the R_m values, the elongations at R_m, e_n (n after necking), and the yield stress, YS, are reported in Table 3. In fact, the comparison of the model flow curves in engineering stress–strain affords an extended evaluation of the model results, since all flow curves strained beyond necking correspond to the end of uniform elongation and the occurrence of localised deformation. Since the final rupture e_R could be affected by local defects in the necking, e_R prediction was beyond the aims of the present investigation.

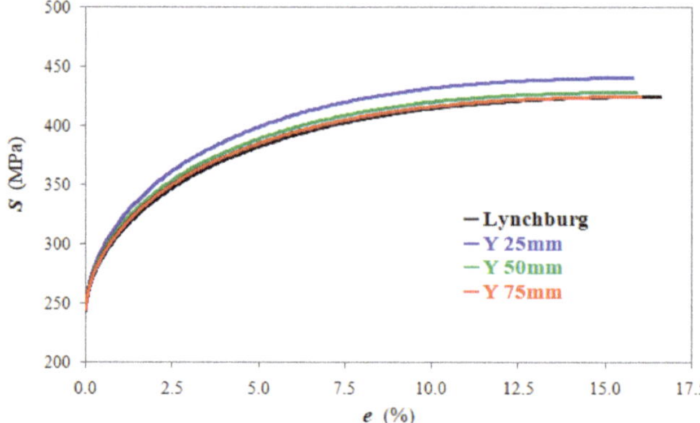

Figure 5. Engineering stress–strain flow curves (up to ultimate tensile stress R_m) built with the average microstructure parameters for the four different moulds reported in Table 2.

Table 3. Comparison between the predicted (engineering flow curves reported in Figure 5) and experimental (^{exp}) average tensile properties, ultimate tensile strength R_m, yield stress YS, and elongation at necking e_n.

Mould	R_m (MPa)	R_m^{exp} (MPa)	e_n (%)	e_n^{exp} (%)	YS (MPa)	YS^{exp} (MPa)
Lynchburg 25 mm	424.4	424.3	16.6	16.7	277.2	288.3
Y 25 mm	440.5	440.7	15.8	16.0	277.9	294.2
Y 50 mm	428.4	429.8	15.9	16.2	278.7	288.8
Y 75 mm	424.5	426.5	16.0	16.0	277.	287.7

The comparison of the model results and the experimental data reported in Table 3 proves that the plastic behaviour of the GJS 400 produced with different cooling rates (different thicknesses) can be described successfully by using the classical strain hardening model widely used for ductile metallic materials [22–28]. Since the correlation between the mechanical constituents (ferrite and pearlite), physical parameters, and microstructure was validated, the use of dislocation-related-dislocation density constitutive equations (like the Voce and Estrin equations) for different DI grades reported in previous investigations [27,28] was also validated. Considerations about the other microstructure parameters are reported in Section 4.3.

Indeed, even if the use of the simple rule of mixture is diffused in DP steels [31–33] that have constituents (ferrite and martensite) that are similar (from the perspective of hardness) to GJS 400 (ferrite and pearlite), the use of the rule of mixture has not been always successful. Particularly when the volume fractions of the constituents have varied widely [33], some modifications to the rule of mixture have been necessary. Indeed, Equation (5) is consistent with an iso-strain approach, but kinematic hardening should occur at the boundaries between ferrite and pearlite during the early stages of straining, and isotropic hardening because of diffuse dislocation activities in the soft ferrite grains should become significant at large strains. However, in this study, the pearlite volume fraction varied slightly and never excided 4%, so this limited range could explain why the simple rule of mixture (Equation (5)) worked well in the present investigation.

The sensitivity of the model to the microstructure parameters was tested by opportunely changing the average microstructure parameters. Examples of this investigation are reported in Figure 6. In Figure 6a, engineering flow curves from the Lynchburg and the Y 25 mm moulds are shown, while in Figure 6b, the pearlite volume fraction in GJS 400 from the Y 25 mm mould was set to zero (like in the Lynchburg mould), resulting in a flow curve lower than the Lynchburg samples with a smaller ferritic grain size. In Figure 6c, the flow curves from the Lynchburg, Y 50 mm, and Y 75 mm moulds are reported, while in Figure 6d, the average pearlite volume fractions from Y 50 mm and Y 75 mm were set to zero (like in Lynchburg), resulting again in flow curves lower than those from the Lynchburg samples with smaller average grain sizes.

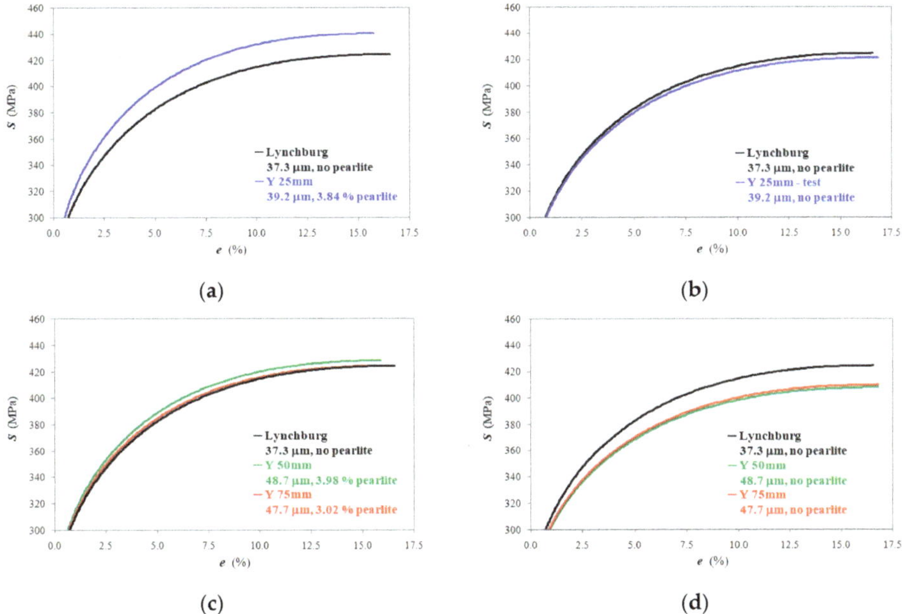

Figure 6. Engineering stress–strain curves (up to an ultimate tensile stress of R_m): (**a**) Lynchburg and Y 25 mm moulds and (**b**) ferrite contribution to flow curves only; (**c**) Lynchburg, Y 50 mm and Y 75 mm moulds; (**d**) ferrite contribution to flow curves only.

From the results reported in Figure 6, it can be concluded that the GJS 400 from Y 25 mm mould was the strongest (R_m = 440.5 MPa) and the least ductile (elongation at R_m = 15.8%) (in Figure 5) because of the combination of its small ferritic grain size (39.2 ± 2.3 μm) and high pearlite volume fraction (3.8 ± 0.4%), while in the Lynchburg mould, even if the ferritic microstructure was the finest (37.3 ± 3.2 μm) because it had the highest solidification rate, the absence of pearlite produced the softest microstructure with the lowest R_m (424.4 MPa) and the most ductile microstructure with the largest elongation at R_m (16.6%). The significant increases of ferritic grain size in the Y 50 mm (48.6 ± 4.7 μm) and Y 75 mm (47.7 ± 7.0 μm) samples, which should have significantly weakened the GJS 400 microstructures, were, indeed, compensated by the significant presence of pearlite (4.0% ± 1.6% in Y 50 mm and 3.0% ± 0.5% in the Y 75 mm mould). Thus, the engineering flow curves in Figure 6c for the Lynchburg and Y 50 mm and 75 mm moulds were finally comparable. It is noteworthy that the model also correctly described the elongations at R_m, since in the comparable flow curves (the Lynchburg, Y 50 mm, and Y 75 mm moulds), the microstructures with the higher pearlite volume fractions presented shorter elongations to R_m (15.9% in Y 50 mm and 16.0% in the Y 75 mm mould), which is consistent with the fact that the microstructure constituents that strengthen materials reduced their ductility.

4.3. Considerations of Other Microstructural Parameters

The spherical shape of graphite confers high ductility to the cast irons (producing so-called Ductile Irons (DIs)), and different graphite volume fractions and graphite morphologies could the affect tensile properties. Improper graphite shapes can give rise to stress-raisers that firstly affect the ductility, and if the nature and the density of the stress-raisers are particularly severe, they could also affect the plastic behaviours and the tensile flow curves [13–16]. In this case, the parameters for the graphitic nodules should be taken into account. However, the average nodularity values (ASTM 2567-16a) of the GJS 400 produced with the different moulds were excellent and over the minimum of 80%, thereby complying with the standard (namely 89.8% ± 3.0% for Lynchburg, 91.2% ± 1.6% for Y 25 mm, 87.1% ± 1.4% for Y 50 mm, and 83.2% ± 4.6% for Y 75 mm. The high nodularity and almost constant graphite volume fractions in the investigated GJS 400 samples could explain why there was no need to involve nodule parameters in the plasticity model of Equations (3)–(5).

According to the literature on steels [33,34] and DIs [7,35], the interlamellar spacing λ of pearlite, which depends on the cooling rate at the critical temperature A_{c1} when the eutectoid transformation starts, should have effect on mechanical properties. In the present GJS 400 investigation [17], the cooling rates at Ac1 were 2.40 °C/min in the Y 75 mm mould, 3.54 °C/min in the Y 50 mm mould, and 5.47 °C/min in Y 25 mm. In the GJS 400 samples produced with different moulds [17], the pearlite was irregular, and its shape was rarely lamellar. It depended instead on grain orientation in agreement with [7,35], so it was not possible to measure any characteristic interlamellar spacing. However, the results reported in Figure 4, where a single pearlite flow curve was valid for all samples, and the fact that the pearlite volume fraction was the only significant parameter, suggest that the pearlite's characteristic widths likely did not change significantly in the range of the investigated cooling rates through A_{c1}. However, the pearlite volume fractions reported in Table 2 varied slightly from 0% to 4% in the different moulds, and this could be another possible reason why a single pearlite flow curve (i.e., a single pearlite characteristic width) could be used successfully.

Thus the microstructure plasticity model allowed to accurately describe the experimental flow curves of the GJS 400 produced with different cooling rates, proving that ferritic grain sizes and pearlite volume fractions mainly affect the plastic flow behaviour of GJS 400 in agreement with the microstructure–mechanical property relations reviewed in [12], demonstrating that the strain hardening behaviour and strength of DIs are strongly affected by their microstructures, since yield stress increases with decreasing ferritic grain size, while in the ferritic–pearlitic Dis, the yield stress increases with an increasing content of pearlite, which causes a reduction of the strain to fracture. Minor mismatching was found at small strains, which could be rationalized in term of the decohesion between the graphite nodules and the ferritic matrix. The void nucleation, caused by graphite-matrix decohesion, followed by void growth and coalescence, can be expected to affect all the tensile flow curves of DJS 400. Detailed investigations [36–38] on the plastic behaviour of a ferritic DI reported that the graphite-matrix decohesion did not cause any dramatic decrease in tensile stress, though the damage was significant with a final failure of about 20%. Thus, the tensile flow curves at high stresses were representative of the microstructures, and the graphite nodules decohesion affected the flow curves only at yielding. In order to take into proper consideration the graphite-matrix decohesion to describe yielding in the DIs, interesting results have been reported via numerical simulations on the effects of residual stresses at the graphite–matrix interface and decohesion at the early stages of deformation, with an estimation of increased *YS* of about 5% [39,40], which is consistent with the mismatch at *YS* reported in Table 3, with errors of *YS* between −3.7% (the Y 75 mm mould) and −5.6% (the Lynchburg mould).

5. Conclusions

Different microstructures of GJS 400 were obtained through different cooling rates. The microstructures were characterised in detail elsewhere [17], quantifying the microstructure's parameters, like nodule count, nodularity, the average diameter of the graphite nodules, volume fractions of the graphite and pearlite (complying with the standard ASTM E2567-16a), and the average ferritic grain size (complying with the standard ASTM E112-96). The tensile flow curves were modelled with the Follansbee and Estrin-Kocks-Mecking approaches, which afforded an explicit correlation between the plastic behaviour and microstructure parameters. The following conclusions were achieved:

- This model described very well the experimental flow curves at high strains, while at low strains, minor mismatching was present. This mismatching was ascribed to the graphite-matrix decohesion;
- The plastic behaviour of the GJS 400 with different microstructures depended mainly on the ferritic grain size and pearlitic volume fraction, while the other microstructure parameters were not needed to rationalize the GJS 400's plastic behaviour;
- The correlation between the mechanical constituents (ferrite and pearlite), physical parameters, and microstructure was validated, so the use of dislocation-related-dislocation density constitutive equations (like the Voce and Estrin equations) for different DI grades reported in previous investigations was also validated;
- The results proved that the data gathered while complying with the minimum requirements of the standards' statistics were not enough to produce accurate microstructural data.

Author Contributions: Conceptualization, G.A.; methodology, G.A.; resources, G.A. and F.Z.; data analysis, G.A., R.D. and D.R.; writing—original draft preparation, G.A.; writing—review and editing, R.D., D.R., and M.G.; supervision, G.A.

Funding: This research received no external funding.

Acknowledgments: Davide Della Torre and Tullio Ranucci are warmly thanked for their experimental support.

Conflicts of Interest: The authors declare no conflict of interest.

Appendix A. Mechanistic Equation of Strain Hardening and Physical Parameters

In the frame of the strain hardening theory by Kocks-Mecking-Estrin [22–28], the total dislocation-density ρ increases because of straining according to the mechanistic equation

$$\frac{d\rho}{d\varepsilon_P} = M \cdot \left[\left(\frac{1}{b\Lambda} + \frac{1}{bD} + \frac{1}{b\lambda}\right) - D_o \cdot \rho\right], \tag{A1}$$

where D_o = dynamic recovery term;

Λ = dislocation mean free path related to the dislocation cells in ferrite;

D = ferritic grain size or pearlitic island size;

λ = interlamellar spacing in pearlite.

According to principle of similitude $\Lambda = \beta/\rho^{1/2}$, with a β constant of the magnitude between 100 and 200 [22], Equation (A1) becomes

$$\frac{d\rho}{d\varepsilon_P} = M \cdot \left[\left(\frac{\sqrt{\rho}}{b\beta} + \frac{1}{bD} + \frac{1}{b\lambda}\right) - D_o \cdot \rho\right], \tag{A2}$$

and

$$2\frac{d\sqrt{\rho}}{d\varepsilon_P} = M \cdot \left[\left(\frac{1}{b\beta} + \frac{1}{bD\sqrt{\rho}} + \frac{1}{b\lambda\sqrt{\rho}}\right) - D_o \cdot \sqrt{\rho}\right], \tag{A3}$$

From Equation (1) in Section 2.2

$$\sqrt{\rho} = \sigma_G/(M\alpha_o Gb), \tag{A4}$$

so Equation (A3) becomes

$$\frac{2}{M\alpha_o Gb}\frac{d\sigma_G}{d\varepsilon_P} = M \cdot \left[\left(\frac{1}{b\beta} + \frac{M\alpha_o Gb}{bD\sigma_G} + \frac{M\alpha_o Gb}{b\lambda\sigma_G}\right) - D_o \cdot \frac{\sigma_G}{M\alpha_o Gb}\right], \tag{A5}$$

and

$$\frac{d\sigma_G}{d\varepsilon_P} = \left[\frac{M^2\alpha_o G}{2\beta} + \frac{M^3(\alpha_o G)^2 b}{2D\sigma_G} + \frac{M^3(\alpha_o G)^2 b}{2\lambda\sigma_G}\right] - \frac{MD_o}{2} \cdot \sigma_G, \tag{A6}$$

Which, if written in a more compact way, results in

$$\frac{\mathrm{d}\sigma_G}{\mathrm{d}\varepsilon_P} = \left[\frac{K_o}{\beta} + \frac{K_1}{D\sigma_G} + \frac{K_1}{\lambda\sigma_G}\right] - \frac{\sigma_G}{\varepsilon_c}, \tag{A7}$$

where M is the Taylor factor (3.01 in BCC materials), α_o the dislocation–dislocation interaction strength (0.5) [22], G the elastic shear modulus for ferrite (64 GPa), and b the Burgers vector length of the ferrite (0.248 nm), which results in

$$K_o = \frac{M^2\alpha_o G}{2} = 1.538 \times 10^5 \text{ MPa}, \tag{A8}$$

and

$$K_1 = \frac{M^3(\alpha_o G)^2 b}{2} = 7.565 \times 10^6 \text{ MPa}^2\cdot\mu m, \tag{A9}$$

as reported in Section 2.2.

References

1. Gonzaga, R.A.; Carrasquilla, J.F. Influence of an appropriate balance of the alloying elements on microstructure and on mechanical properties of nodular cast iron. *J. Mater. Process. Tech.* **2005**, *162*, 293–297. [CrossRef]
2. Gonzaga, R.A. Influence of ferrite and pearlite content on mechanical properties of ductile cast irons. *Mater. Sci. Eng. A* **2013**, *567*, 1–8. [CrossRef]
3. Alhussein, A.; Risbet, M.; Bastien, A.; Chobaut, J.P.; Balloy, D.; Favergeon, J. Influence of silicon and addition elements on the mechanical behaviour of ferritic ductile cast iron. *Mater. Sci. Eng. A* **2014**, *605*, 222–228. [CrossRef]
4. Bradley, W.L.; Srinivasan, M.N. Fracture and fracture toughness of cast irons. *Int. Mater. Rev.* **1990**, *35*, 129–161. [CrossRef]
5. Iacoviello, F.; Di Bartolomeo, O.; Di Cocco, V.; Piacente, V. Damaging micro-mechanisms in ferritic-pearlitic ductile cast irons. *Mater. Sci. Eng. A* **2008**, *478*, 181–186. [CrossRef]
6. Di Cocco, V.; Iacoviello, F.; Rossi, A.; Iacoviello, D. Macro and microscopical approach to the damaging mechanisms analysis in a ferritic ductile cast iron. *Theor. Appl. Fract. Mech.* **2014**, *69*, 26–33. [CrossRef]
7. Guo, X.; Stefanescu, D.M.; Chuzhoy, L.; Pershing, M.A.; Biltgen, G.L. A Mechanical Properties Model for Ductile Iron. *AFS Trans.* **1997**, *105*, 47–54.
8. Donelan, P. Modelling microstructural and mechanical properties of ferritic ductile cast iron. *Mater. Sci. Tech. Ser.* **2000**, *16*, 261–269. [CrossRef]
9. Kasvayee, K.A.; Ghassemali, E.; Svensson, I.L.; Olofsson, J.; Jarfors, A.E.W. Characterization and modeling of the mechanical behavior of high silicon ductile iron. *Mater. Sci. Eng. A* **2017**, *708*, 159–170. [CrossRef]
10. Svensson, I.L.; Sjögren, T. On modeling and simulation of mechanical properties of cast irons with different morphologies of graphite. *Int. J. Met.* **2009**, *3*, 67–77. [CrossRef]
11. Olofsson, J.; Salomonsson, K.; Svensson, I.L. Modelling and simulations of ductile iron solidification induced variations in mechanical behaviour on component and microstructural level. *IOP Conf. Ser. Mater. Sci. Eng.* **2015**, *84*, 012026. [CrossRef]
12. Hütter, G.; Zybell, L.; Kuna, M. Micromechanisms of fracture in nodular cast iron: From experimental findings towards modeling strategies—A review. *Eng. Fract. Mech.* **2015**, *144*, 118–141. [CrossRef]
13. Goodrich, G.M. Cast iron microstructure anomalies and their causes. *AFS Trans.* **1997**, *105*, 669–683.
14. Iwabuchi, Y.; Narita, H.; Tsumura, O. Toughness and Ductility of heavy-walled ferritic spheroidal-graphite iron castings. *Res. Rep. Kushiro Natl. Coll.* **2003**, *37*, 1–9.
15. Nilsson, K.F.; Blagoeva, D.; Moretto, P. An experimental and numerical analysis to correlate variation in ductility to defects and microstructure in ductile cast iron components. *Eng. Fract. Mech.* **2006**, *73*, 1133–1157. [CrossRef]
16. Nilsson, K.F.; Vokal, V. Analysis of ductile cast iron tensile tests to relate ductility variation to casting defects and material microstructure. *Mater. Sci. Eng. A* **2009**, *502*, 54–63. [CrossRef]

17. Angella, G.; Ripamonti, D.; Górny, M.; Masaggia, S.; Zanardi, F. The role of microstructure on tensile plastic behaviour of ductile iron GJS 400 produced through different cooling rates—Part I: Microstructure. *Metals* **2019**, in press.
18. Rivera, G.; Boeri, R.; Sikora, J. Revealing the solidification structure of nodular iron. *Int. J. Cast Metal. Res.* **1995**, *8*, 1–5. [CrossRef]
19. Rivera, G.; Boeri, R.; Sikora, J. Revealing and characterizing solidification structure of ductile cast iron. *Mater. Sci. Technol.* **2002**, *18*, 691–698. [CrossRef]
20. Follansbee, P.S.; Kocks, U.F. A Constitutive Description of the Deformation of Copper Based on the Use of the Mechanical Threshold Stress as an Internal State Variable. *Acta Metall. Mater.* **1988**, *36*, 81–93. [CrossRef]
21. Follansbee, P.S. Structure Evolution in Austenitic Stainless Steels -A State Variable Model Assessment. *Mater. Sci. Appl.* **2015**, *6*, 457–463. [CrossRef]
22. Kocks, U.F.; Mecking, H. Physics and phenomenology of strain hardening: The FCC case. *Prog. Mater. Sci.* **2003**, *48*, 171–273. [CrossRef]
23. Estrin, Y.; Mecking, H. A unified phenomenological description of work hardening and creep based on one-parameter models. *Acta Metall. Mater.* **1984**, *32*, 57–70. [CrossRef]
24. Estrin, Y. Dislocation theory based constitutive modelling: Foundations and applications. *J. Mater. Process. Tech.* **1998**, *80*, 33–39. [CrossRef]
25. Angella, G. Strain hardening analysis of an austenitic stainless steel at high temperatures based on the one-parameter model. *Mater. Sci. Eng. A* **2012**, *532*, 381–391. [CrossRef]
26. Angella, G.; Donnini, R.; Maldini, M.; Ripamonti, D. Combination between Voce formalism and improved Kocks-Mecking approach to model small strains of flow curves at high temperatures. *Mater. Sci. Eng. A* **2014**, *594*, 381–388. [CrossRef]
27. Angella, G.; Zanardi, F.; Donnini, R. On the significance to use dislocation-density-related constitutive equations to correlate strain hardening with microstructure of metallic alloys: The case of conventional and austempered ductile irons. *J. Alloy. Compd.* **2016**, *669*, 262–271. [CrossRef]
28. Donnini, R.; Fabrizi, A.; Bonollo, F.; Zanardi, F.; Angella, G. Assessment of the Microstructure Evolution of an Austempered Ductile Iron During Austempering Process Through Strain Hardening Analysis. *Met. Mater. Int.* **2017**, *23*, 855–864. [CrossRef]
29. Caballero, F.G.; Garcìa de Andrés, C.; Capdevila, C. Characterization and morphological analysis of pearlite in a eutectoid steel. *Mater. Charact.* **2000**, *45*, 111–116. [CrossRef]
30. Dieter, G.E. *Mechanical Metallurgy*; McGraw-Hill Book Company: New York, NY, USA, 1988; pp. 208–212.
31. Byun, T.S.; Kim, I.S. Tensile properties and inhomogeneous deformation of ferrite-martensite dual phase steels. *J. Mater. Sci.* **1998**, *9*, 85–92. [CrossRef]
32. Park, K.; Nishiyama, M.; Nakada, N.; Tsuchiyama, T.; Takaki, S. Effect of the martensite distribution on the strain hardening and ductile fracture behaviours of dual phase steel. *Mater. Sci. Eng. A* **2014**, *604*, 135–141. [CrossRef]
33. Alibeyki, M.; Mirzadeh, H.; Najafi, M.; Kalhor, A. Modification of rule of mixtures for estimation of the mechanical properties of dual phase steels. *J. Mater. Eng. Perform.* **2017**, *26*, 2683–2688. [CrossRef]
34. Modi, O.P.; Deshmukh, N.; Mondal, D.P.; Jha, A.K.; Yegneswaran, A.H.; Khair, H.K. Effect of interlamellar spacing on the mechanical properties of 0.65% C steel. *Mater. Charact.* **2001**, *46*, 347–352. [CrossRef]
35. Benavides-Treviño, J.R.; Perez-Gonzalez, F.A.; Hernandez, M.A.L.; Garcia, E.; Juarez-Hernandez, A. Influence of the cooling rate on the amount of graphite nodules and interlamellar space in pearlite phase in a ductile iron. *Indian J. Eng. Mater. Sci.* **2018**, *25*, 330–334.
36. Dong, M.J.; Prioul, C.; Francois, D. Damage effect on the fracture toughness of nodular cast iron: Part I. damage characterization and plastic flow stress modelling. *Metall. Mater. Trans. A* **1997**, *28*, 2245–2254. [CrossRef]
37. Guillemer-Neel, C.; Feaugas, X.; Clavel, M. Mechanical behavior and damage kinetics in nodular cast iron: Part I. Damage mechanisms. *Metall. Mater. Trans. A* **2000**, *31*, 3063–3074. [CrossRef]
38. Guillemer-Neel, C.; Feaugas, X.; Clavel, M. Mechanical behavior and damage kinetics in nodular cast iron: Part II. Hardening and damage. *Metall. Mater. Trans. A* **2000**, *31*, 3075–3086. [CrossRef]

39. Zhang, Y.B.; Andriollo, T.; Fæster, S.; Liu, W.; Hattel, J.; Barabash, R.I. Three-dimensional local residual stress and orientation gradients near graphite nodules in ductile cast iron. *Acta Metall. Mater.* **2016**, *121*, 173–180. [CrossRef]
40. Andriollo, T.; Zhang, Y.; Fæster, S.; Thorborg, J.; Hattel, J. Impact of micro-scale residual stress on in-situ tensile testing of ductile cast iron: Digital volume correlation vs. model with fully resolved microstructure vs. periodic unit cell. *J. Mech. Phys. Solids* **2019**, *125*, 714–735. [CrossRef]

Article

Strength Prediction for Pearlitic Lamellar Graphite Iron: Model Validation

Vasilios Fourlakidis [1],*, Ilia Belov [2] and Attila Diószegi [2]

[1]   Swerea SWECAST, 550 02 Jönköping, Sweden
[2]   Materials and Manufacturing-Foundry Technology, Jönköping University, Gjuterigatan 5,
     553 18 Jönköping, Sweden; ilia.belov@ju.se (I.B.); attila.dioszegi@ju.se (A.D.)
*   Correspondence: vasilios.fourlakidis@swerea.se; Tel.: +46-036-30-12-07

Received: 11 July 2018; Accepted: 30 August 2018; Published: 31 August 2018

Abstract: The present work provides validation of the ultimate tensile strength computational models, based on full-scale lamellar graphite iron casting process simulation, against previously obtained experimental data. Microstructure models have been combined with modified Griffith and Hall–Petch equations, and incorporated into casting simulation software, to enable the strength prediction for four pearlitic lamellar cast iron alloys with various carbon contents. The results show that the developed models can be successfully applied within the strength prediction methodology along with the simulation tools, for a wide range of carbon contents and for different solidification rates typical for both thin- and thick-walled complex-shaped iron castings.

Keywords: lamellar graphite iron; ultimate tensile strength; primary austenite; gravity casting process simulation

---

1. Introduction

Nowadays, there is a great need to further improve both the material properties and the prediction models for optimization of the heavy truck engine components aimed to fulfil the rigorous environmental legislations, sustainability goals, and customer demands. Cylinder blocks and cylinder heads are the primary components of these engines, and the majority of them are composed of lamellar graphite iron (LGI). The ultimate tensile strength (UTS) of LGI is an essential material property that determines the engine performance and the fuel consumption. The complex geometry and variation of the wall thickness in the cylinder blocks result in different solidification times through the component, and thus, different tensile properties.

A number of investigators [1–6] underlined the major influence of the graphite flake size on the strength of LGI. It is believed that under stress, the graphite flakes are dispersed in the metal matrix act as notches that decrease the material strength. Modified Griffith and Hall–Petch models were introduced for the prediction of UTS in LGI, where the maximum graphite length was considered as the maximum defect size [3,7–9]. Recently, it was found that the maximum defect size can never be larger than the interdendritic space between the primary austenite dendrites formed during the solidification process [10]. The length scale of the interdendritic space was characterized by the hydraulic diameter of the interdendritic phase (D_{IP}^{Hyd}), which proved to be the most suitable parameter to express the detrimental effect of the graphite lamella in the metallic matrix. Thus, the D_{IP}^{Hyd} parameter was introduced as the maximum defect size in the modified Griffith and Hall–Petch equations [10,11].

Over the past decades, computer simulations of LGI solidification were carried out by several researchers [7–9,12,13] to describe the thermal history and the microstructure evolution of LGI castings. The main objective of these studies was prediction of the UTS. Macroscopic heat flow modeling, coupled with growth kinetic equations, was introduced in [7] to predict various microstructure features of LGI. Consequently, a modified Griffith fracture relation was applied to determine the UTS of a commercial LGI alloy. A similar solidification model was developed in [8], where a microstructure evolution model was employed together with the modified Hall–Petch equation for calculation of the UTS. Note that in [8], two different cooling rates resulted in two different relationships between the UTS and the maximum graphite flake length. Similar observations were made in [10], where three different cooling rates led to providing three different linear dependencies between the eutectic cell size (direct proportional to the maximum graphite length) and the UTS.

The present work provides validation of the UTS computational models against experimental data, based on full-scale pearlitic LGI gravity casting process simulation. We investigated whether the models recently developed in [10,11] can be applied within the UTS prediction methodology, along with the simulation tools, for different alloy compositions and for different solidification rates. The novel methodology for UTS prediction, presented in this paper, involves D_{IP}^{Hyd} as the key morphological parameter, along with the pearlite lamellar spacing. These parameters are dependent on solidification time, cooling rate, and alloy composition. The proposed approach bears simplicity compared to the microstructure modelling methods [7,8]. The methodology is validated to include analytical formulation of the UTS prediction models and robust experimental thermal analysis, to obtain latent heat of solidification and solid-state transformation as input data for the simulation. First, the UTS modeling methods are elaborated followed by the details on the experimental setup and alloy composition. Casting simulation model is then introduced, as well as the simulation procedure. The results are discussed in comparison with the temperature and UTS measurements, followed by conclusions regarding applicability and limitations of the proposed UTS prediction methodology.

2. UTS Modeling

The modified Griffith fracture relation is given by Equation (1) [3], and the modified Hall–Petch strengthening model is represented by Equation (2) [8].

$$\sigma_{UTS} = \frac{k_t}{\sqrt{\alpha}} \tag{1}$$

$$\sigma_{UTS} = k_1 + \frac{k_2}{\sqrt{d}}, \tag{2}$$

where σ_{UTS} is the ultimate tensile strength, α is the maximum defect size, and k_t is the stress intensity factor of the metallic matrix, k_1 and k_2 are the contributions from other strengthening mechanisms, and d is the grain size. The maximum defect size and grain size, α and d, are provided in μm, parameters k_t and k_2 are in MPa, $\sqrt{\mu m}$, and k_1 is in MPa.

It was found in [10] that D_{IP}^{Hyd} is the dominant factor that reduces the UTS in lamellar graphite iron alloys. A modified Griffith equation was obtained in [10] as result of the linear regression analysis of the experimental data, Equation (3).

$$\sigma_{UTS} = \frac{1212}{\sqrt{D_{IP}^{Hyd}}} \tag{3}$$

According to this model, if a tensile force is applied on the microstructure, a crack will start to form at a certain stress level. The crack will propagate relatively easily through the numerous interconnected graphite particles that are embedded in the metallic matrix of the eutectic cell. When the crack reaches the metallic matrix (pearlite) that was originated from the primary austenite (dendritic phase), the relatively rapid crack extension will be halted, due to the fact that much larger stresses are required for the fracture of this phase. The magnitude of the additional stress is proportional to the pearlite lamellar spacing ($\lambda_{pearlite}$). Based on this assumption, it becomes apparent that the effect of $\lambda_{pearlite}$ on the UTS must be taken into consideration. Thus, linear multiple regression analysis was made to determine the simultaneous influence of the D_{IP}^{Hyd} and the $\lambda_{pearlite}$ on the UTS. The model obtained is based on the modified Hall–Petch relation, and is expressed by Equation (4) [11].

$$\sigma_{UTS} = 70.9 + \frac{491.2}{\sqrt{D_{IP}^{Hyd}}} + \frac{295.7}{\sqrt{D_{IP}^{Hyd} \cdot \lambda_{pearlite}}} \tag{4}$$

The D_{IP}^{Hyd} parameter was found to be related to the solidification time (t_s) and the fraction of primary austenite (f_γ), as seen from Equation (5) [14].

$$D_{IP}^{Hyd} = \frac{1}{0.8 \cdot f_\gamma} \cdot t_s^{\frac{1}{3}} \tag{5}$$

The $\lambda_{pearlite}$ parameter at room temperature was assumed to be dependent on the cooling rate in the eutectoid transformation region. The empirical relationship between $\lambda_{pearlite}$ at room temperature, and the cooling rate at the temperature intervals between 700 and 740 °C, is shown in Figure 1. The experimentally derived relation Equation (6) was used for investigating the effect of different $\lambda_{pearlite}$ prediction models on simulated UTS. The measurements techniques, the microstructure and thermal data that resulted in Equation (6), are presented elsewhere [11,12]. Briefly, the pearlite lamellar spacing was measured using SEM and a linear intercept method. The minimum value was considered to be the correct spacing (perpendicular to the lamellae). The distance between 11 adjacent ferrite lamellas was measured and divided by 10 for estimation of a single interlamellar spacing.

$$\lambda_{pearlite} = 0.054 \cdot \left(\frac{dT}{dt}_{[700;740\ °C]} \right)^{-0.525} \tag{6}$$

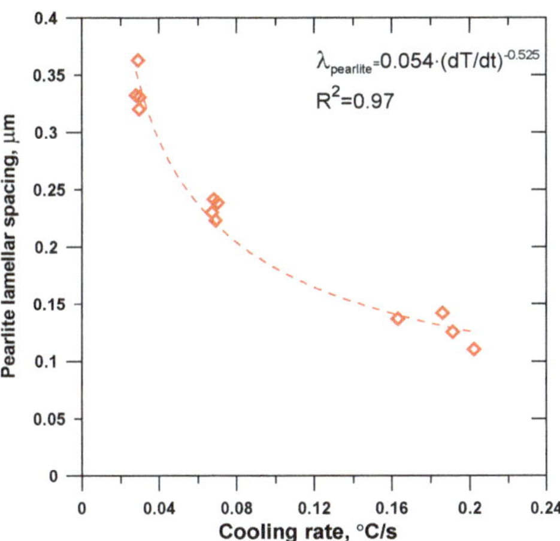

Figure 1. Pearlite lamellar spacing as function of cooling rate between 700 and 740 °C.

3. Materials and Methods

3.1. Cylindrical Castings

The experimental layout contained three cylindrical cavities, each one surrounded by a different material (steel chill, sand, and insulation) intended to provide three different cooling rates. The entire assembly was enclosed by a furan-bounded sand mold. The dimensions of the cylinders surrounded by sand and chill were ⌀50 × 70 mm, and the insulated cylinder dimensions were ⌀80 × 70 mm.

A lateral 2-D heat flow condition was induced by placing an insulation plate at the top and bottom of the cylindrical castings. The design of the cylindrical castings and arrangement of the experimental layout are shown in Figures 2 and 3, respectively.

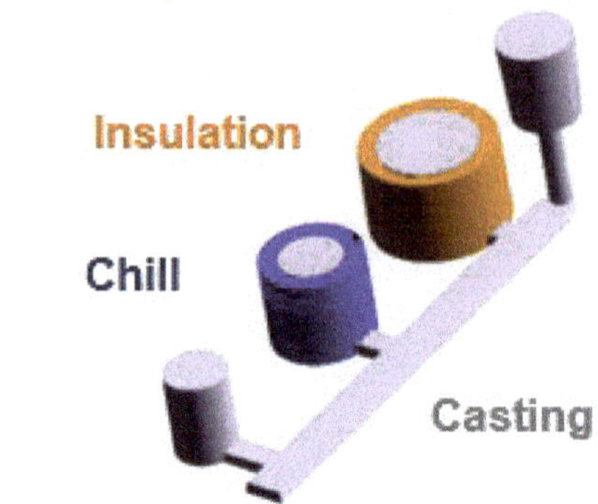

Figure 2. Cylindrical castings with the insulation and chill.

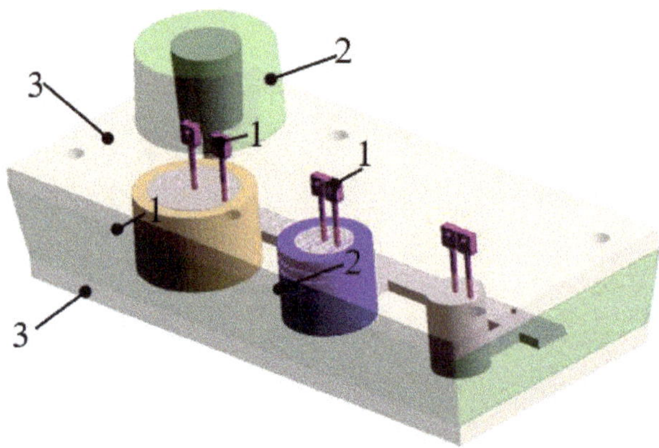

Figure 3. The experimental layout. (1) Thermocouples, (2) sand mold, and (3) insulation plates.

Two type S thermocouples with glass tube protection were embedded in every cylindrical casting. A central thermocouple was located on the central axis of the cylinder. The distance between the central and the lateral thermocouple was 20 mm for the $\varnothing$50 mm cylinder and 30 mm for the $\varnothing$80 mm cylinder. The thermocouples were placed at the mid-height of each cylinder and the temperatures were recorded at approximately 0.2 s interval. A 16-bit resolution data acquisition system with the sampling rate 100 Hz was employed [12].

The mold-filling time was 12 s. The solidification times of the metal in the chill, sand, and insulation were roughly 80, 400, and 1500 s, respectively. An electric induction furnace was utilized for melting of the charge material. The cast iron base alloy was inoculated with a constant level of a standard Sr-based inoculant. Four hypoeutectic lamellar graphite iron heats with varying carbon contents were produced. The alloy with the higher carbon content was cast first, and steel scraps were added to the furnace for the adjustment of the carbon content in the following casting. Coin-shaped specimens were extracted for chemical analysis. The chemical compositions of the four different alloys are presented in Table 1. All the castings had a fully pearlitic microstructure.

Tensile strength measurements were performed using a dog bone-shaped specimen with 6 mm diameter in the gauge section, 35 mm gauge length, and a 3 μm surface finish. The tests were conducted at a strain rate of 0.035 mm/s and at room temperature. The experimental tensile samples were machined at the distance ~10 mm (sand, chill) and ~20 mm (insulation) from the cylinder axis. The load cell error of the tensile testing machine was <0.5%.

Table 1. Chemical composition (wt %) and carbon equivalent (Ceq = %C + %Si/3 + %P/3).

Alloy	C	Si	Mn	P	S	Cr	Cu	Ceq
A	3.62	1.88	0.57	0.04	0.08	0.14	0.38	4.26
B	3.34	1.83	0.56	0.04	0.08	0.15	0.37	3.96
C	3.05	1.77	0.54	0.04	0.08	0.14	0.36	3.65
D	2.80	1.75	0.54	0.04	0.08	0.15	0.35	3.40

3.2. Simulation Model and Assumptions

A CFD software (Flow-3D CAST, v.5.0 from Flow Science, Inc., Santa Fe, NM, USA) [15] was employed to develop a full-scale 3D model of the casting process for the experimental layout. Mold filling and the cooling/solidification stages were simulated, and local UTS computations were performed on the customized models. Mold-filling time was 12 s, and the laminar flow model was applied. The casting temperature was 1360 °C, and the metal input diameter was 3 cm. The ambient temperature was set to 20 °C. Symmetry boundary conditions were used on the faces of the computational domain, except for the upper face, where the pressure boundary condition was applied. A computational grid of cubical control elements was generated with the cell size 3 mm. The computational grid had a total of ~1 million cells. Different grid densities were tested, and grid-independent results were obtained. The explicit solver was employed during the mold filling, whereas the implicit solver was used for heat transfer simulation in the solidification phase. Since the focus was on heat transfer and the UTS computation methodology, shrinkage and micro-porosity models were not included in the solidification phase.

In this work, the amount of latent heat release due to solidification was related to the solid fraction curves, seen in Figure 4, for the studied alloys. These curves were calculated from the registered experimental cooling curves by using the Fourier thermal analysis method [16,17]. The latent heat of solidification was considered equal to 240 kJ/kg for all studied alloys [18]. Fourier thermal analysis was also applied on cooling curves for the determination of the latent heat release during the eutectoid transformation. The latent heat releases at the eutectoid transformation was found to be similar for all alloys and were incorporated into the specific heat curve as it is shown in Figure 5. The temperature dependent cast iron thermophysical properties [12], and the calibrated heat transfer coefficients applied in the simulation are presented in Table 2.

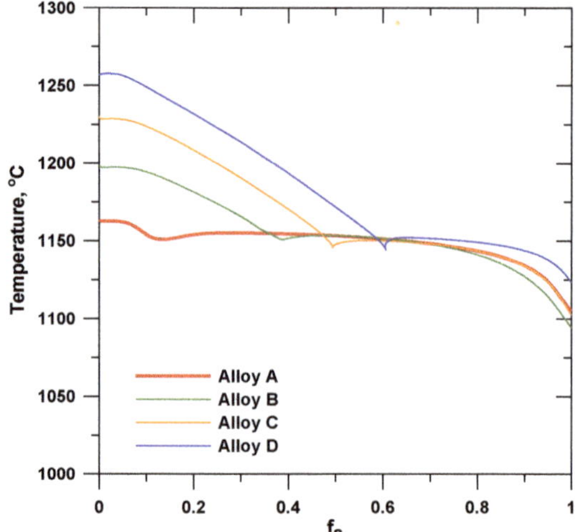

Figure 4. Solid fraction variation with temperature.

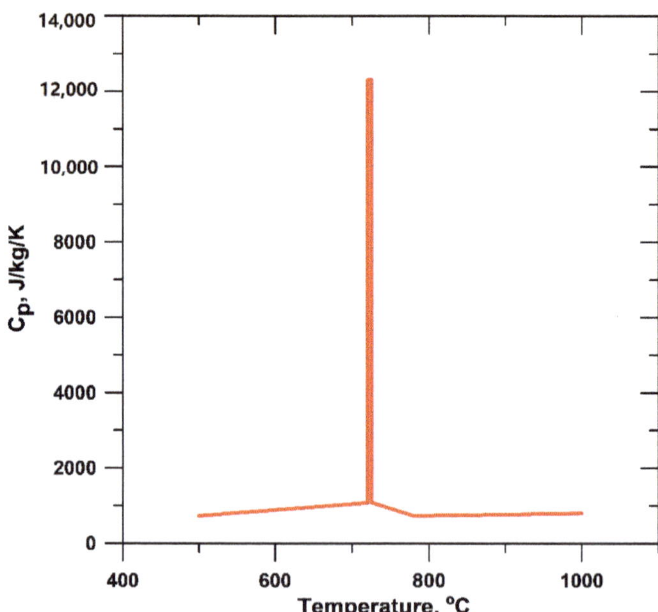

Figure 5. Specific heat as function of temperature.

Table 2. Temperature dependent properties of the cast iron and heat transfer coefficients *.

Temperature (°C)	Cast Iron Thermophysical Properties			Heat Transfer Coefficient		
	Density	Specific Heat	Thermal Conductivity	Sand-Casting	Chill-Casting	Insulation-Casting
	[kg/m^3]	[J/kg/K]	[W/m/K]	[W/m^2/K]	[W/m^2/K]	[W/m^2/K]
600	7146	700	40	40	100	10
720	-	1074	-	-	300	-
721	-	12301	-	-	-	-
724	-	12308	-	-	-	-
725	-	1082	-	50	-	10
750	-	733	-	-	-	-
900	-	-	-	80	-	15
1000	6994	800	-	150	-	25
1100	-	825	-	250	1300	55
1154	6960	837	40	-	1450	-
1170	7016	-	-	-	-	-
1200	6985	-	-	-	1600	60
1227	6939	749	-	-	-	-
1300	6876	771	-	380	-	180
1700	6395	807	38	940	2700	940

* Piecewise linear interpolation was made between neighboring points in the table.

3.3. Simulation Procedure

The simulation procedure consisted of model calibration with respect to the experimental cooling curves available at the location of the central thermocouple. Correct reproduction of the experimental cooling curves is the key for the UTS computation methodology, and one is free to choose methods for model calibration. In this work, the calibration was done by adjustment of the typical heat transfer coefficients between the metal and the insulation, sand, and chill. The UTS calculations for the cylinders were performed during post-processing, by applying local solidification times, local cooling rates in the eutectoid transformation region, and the experimentally determined fraction of primary austenite (f_γ) for each alloy: 0.3 for alloy A, 0.4 for alloy B, 0.51 for alloy C, and 0.61 for alloy D [16].

4. Results and Discussion

The general agreement within 7% was achieved between the simulated and measured cooling curves for insulation-, sand-, and chill-encapsulated cylinders; see Figures 6–9. The larger differences were observed in the solidification region of the chill castings where the eutectic reaction was predicted at higher temperature than measured. This is because the solid fraction-temperature curves were derived from the sand-casting thermal histories, where the undercooling was much lower. Moreover, the solidification model in the simulation used the enthalpy method [19] and ignored the kinetics of phase transformation and, therefore, the undercooling and recalescence of solidification were not predicted. However, the simulated solidification times were in good agreement with the experiment.

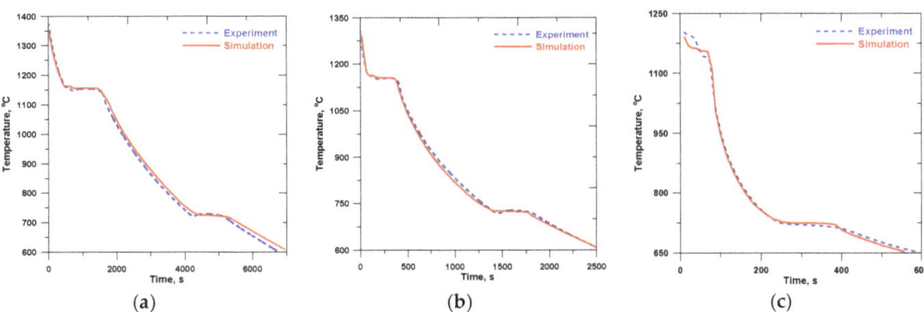

Figure 6. Simulated and experimental cooling curves (central thermocouple) for alloy A: (**a**) insulation, (**b**) sand, and (**c**) chill.

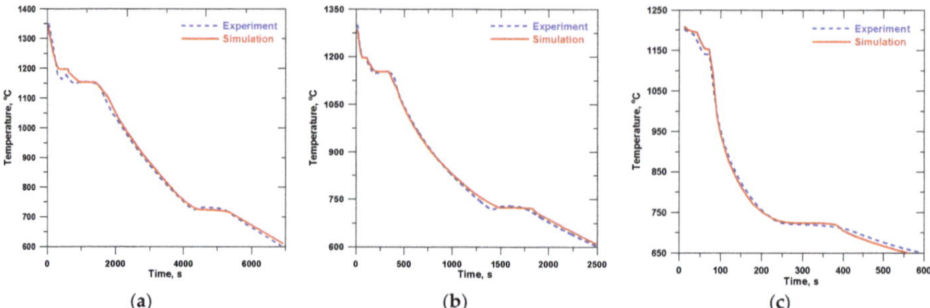

Figure 7. Simulated and experimental cooling curves (central thermocouple) for alloy B: (**a**) insulation, (**b**) sand, and (**c**) chill.

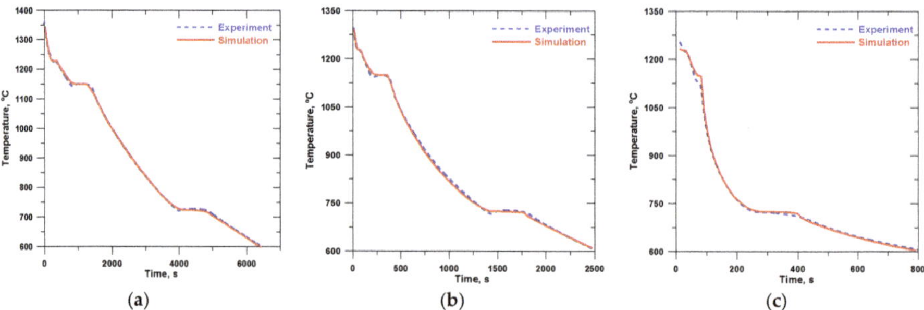

Figure 8. Simulated and experimental cooling curves (central thermocouple) for alloy C: (**a**) insulation, (**b**) sand, and (**c**) chill.

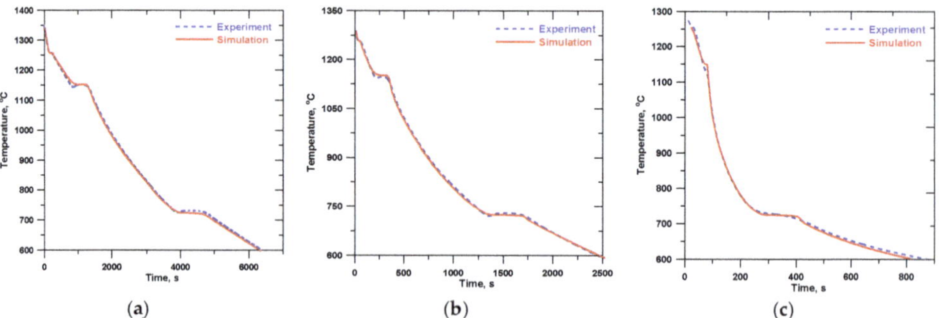

Figure 9. Simulated and experimental cooling curves (central thermocouple) for alloy D: (**a**) insulation (**b**) sand, and (**c**) chill.

The measurement accuracy of the type S thermocouples was ±1.5 °C. It is worth noting that some of the thermocouples inserted in the melt could be slightly displaced from their intended positions during the solidification, which created an additional source of the measurement error; this can be seen clearly, e.g., from the solidification part of the experimental cooling curve for the insulated cylinder in Figure 7.

The simulated solidification times and cooling rates were used in Equations (3) and (4) for the calculation of UTS. The predicted UTS distribution, substituted in the middle cross-section of the alloy B casting, is shown in Figure 10. The figure illustrates the inhomogeneous material strength in the casting.

It is directly related to the temperature gradient and the cooling rate distribution during solidification and solid-state transformation. The reduced UTS is the result of the microstructure coarseness that is related to the solidification time and the cooling rate. Moreover, large UTS gradients on the chilled cylinder can be explained by the large temperature gradients at high solidification rate. Intermediate and slow solidification rates on sand- and insulation-encapsulated cylinders resulted in more uniform distribution of UTS values, due to the smaller temperature gradients during solidification. It should be noted that the variation of UTS magnitude within the tensile bar positions (shown with dashed lines) complicates the model validation.

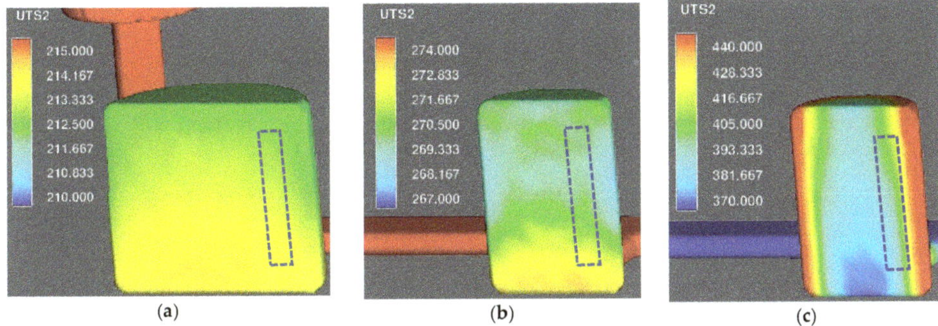

Figure 10. Distribution of ultimate tensile strength (UTS) calculated from Equations (4) and (6) for alloy B: (**a**) insulation-, (**b**) sand-, and (**c**) chill-encapsulated cylinder; the dashed lines indicate the position of the tensile bars.

The obtained values were compared to the measured UTS. Table 3 presents the experimental and simulated UTS results for different cooling rates and for each alloy. The simulated UTS values in Table 3 were picked from the mid-height locations of the tensile bar regions, indicated in Figure 10 with dashed lines. This would correspond to the failure location in the tensile test. However, the exact fracture location might be influenced by several other factors, such as microporosities, graphite flakes that are in contact with the casting surface, or other casting impurities. All of these can cause the crack initiation at positions where the theoretical material strength is not the lowest. Apparently, the fracture analysis is out of scope of the present work. There are quite small differences between simulated and measured UTS values, with the exception of the intermediate and slow cooling rates (sand and insulation) for alloy A, where all the models predicted the UTS with less accuracy. Relatively high, but still acceptable average percentage errors are also observed for the insulated cylinders cast of alloys C and D.

Table 3. Experimental and simulated UTS.

Alloy		UTS, [MPa]			Average Percentage Error, [%]	
		Experiment	Simulation			
			Equation (3) [1]	Equation (4) [2]	Equation (3) [1]	Equation (4) [2]
A	Insulation	154	180	200	17	30
	Sand	195	230	250	18	28
	Chill	363	340–350	340–350	5	5
B	Insulation	211	204	213	3	1
	Sand	254	255	269	1	6
	Chill	368	365–375	385–395	1	6
C	Insulation	250	233	236	7	6
	Sand	286	293	300	2	5
	Chill	440	420–435	435–445	3	0
D	Insulation	289	260	253	10	12
	Sand	337	325	323	4	4
	Chill	447	440–455	475–490	0	8

[1] Modified Griffith model; [2] Modified Hall–Petch model.

Comparisons between the calculated and the measured data are demonstrated in Figure 11. The graph reveals a relatively strong correlation between the measured and computed UTS. The R^2 values show that Equation (3) predicts the UTS with better accuracy than Equation (4). This indicates the need to develop further the model for prediction of the $\lambda_{pearlite}$ parameter.

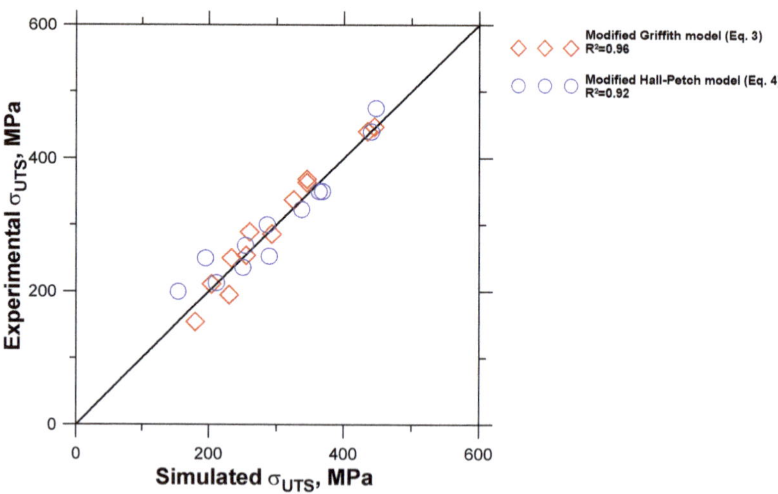

Figure 11. Correlation between measured and simulated UTS values.

The observed deviations between the simulated and measured UTS can be also attributed to the limited number of tensile specimens [10] and to uncertainties regarding the measurements accuracy of the D_{IP}^{Hyd} parameter, especially for the low cooling rate samples [20] that were used to develop the UTS models.

The presented results should be related to two fundamental publications on computer simulations of LGI solidification coupled with the Griffiths and Hall–Petch models [7,8]. The models for UTS calculation utilized in these works were based on a narrow carbon content interval, and on a limited cooling rate variation, in comparison. Moreover, growth kinetic equations were employed in [7,8]. On the contrary, the latent heat release model by the "enthalpy method" [19] was adopted for the solidification simulation in the present work. Furthermore, the presented way to determine the key parameters and incorporate them into material property prediction is novel. In [7,8], the key parameter was the eutectic cell diameter. It is evident that the modified Griffith and Hall–Petch equations are applicable once the eutectic diameter can be predicted, as well as the pearlite lamellar spacing in the Hall–Patch equation. A completely different approach validated in this work involved the hydraulic diameter as the key morphological parameter, along with the pearlite lamellar spacing introduced in [8]. The presented methodology to calculate the UTS features the simplicity of determining the key parameters by simulation (solidification time, cooling rate, and composition dependent). While [7] and [8] introduce complex microstructure models valid for small process intervals (with respect to composition and cooling condition), the current methodology lays back to a robust experimental thermal analysis [16], providing accurate input data (latent heat of both solidification and solid-state transformation) for the simulation. A robust iteration process for tuning up the heat transfer coefficient results in the accurately predicted cooling rate.

5. Conclusions

The novel UTS prediction methodology for fully pearlitic LGI alloys presented in this paper involves hydraulic diameter as the key morphological parameter, along with the pearlite lamellar spacing. It is characterized by simplicity, in comparison to the microstructure modelling methods. The methodology includes analytical formulation of the UTS prediction models, and robust experimental thermal analysis. The latter provides the latent heat of solidification and solid-state transformation as input data for the solidification simulation. In turn, the simulation delivers the solidification time and cooling rates for the UTS prediction models.

Microstructure models for the prediction of hydraulic diameter and the pearlite lamellar spacing, combined with modified Griffith and Hall–Petch equations, were incorporated into casting simulation software for the prediction of UTS in fully pearlitic LGI alloys. Overall, the simulation UTS results were found to be in good agreement (within 9% on the average) with the measurements. However, high average percentage errors were observed for the intermediate and slow cooling rates (sand and insulation) for the alloy with the higher carbon content (alloy A). This study revealed the necessity for development of a more advanced model for the prediction of the $\lambda_{pearlite}$ parameter. The results demonstrated the applicability of the novel UTS prediction models for different chemical compositions and cooling conditions.

Further development of the microstructure modelling would enable determination of the key parameters (hydraulic diameter and pearlite lamellar spacing). However, it seems not to be critical for the presented novel UTS prediction methodology which is valid for the wide process interval.

Author Contributions: A.D. designed the experiment and supervised the work, I.B. performed the simulations, V.F. analyzed the data and wrote the paper, A.D. and I.B. reviewed the paper.

Funding: This research received no external funding.

Acknowledgments: This work was performed within the Swedish Casting Innovation Centre. Cooperating parties are Jönköping University, Scania CV AB, Swerea SWECAST AB and Volvo Powertrain Production Gjuteriet AB. Participating persons from these institutions/companies are acknowledged.

Conflicts of Interest: The authors declare no conflict of interest.

References

1. Collini, L.; Nicoletto, G.; Konecná, R. Microstrucure and mechanical properties of pearlitic lamellar cast iron. *Mater. Sci. Eng. A* **2008**, *488*, 529–539. [CrossRef]
2. Ruff, G.F.; Wallace, J.F. Effects of solidification structure on the tensile properties of lamellar iron. *AFS Trans.* **1977**, *56*, 179–202.

3. Bates, C. Alloy element effect on lamellar iron properties: Part II. *AFS Trans.* **1986**, *94*, 889–905.
4. Nakae, H.; Shin, H. Effect of graphite morphology on tensile properties of flake graphite cast iron. *Mater. Trans.* **2001**, *42*, 1428–1434. [CrossRef]
5. Baker, T.J. The fracture resistance of the flake graphite cast iron. *Mater. Eng. Appl.* **1978**, *1*, 13–18. [CrossRef]
6. Griffin, J.A.; Bates, C.E. Predicting in-situ lamellar cast iron properties: Effects of the pouring temperature and manganese and sulfur concentration. *AFS Trans.* **1988**, *88*, 481–496.
7. Goettsch, D.D.; Dantzig, J.A. Modeling microstructure development in gray cast irons. *Metall. Trans. A* **1994**, *25*, 1063–1079. [CrossRef]
8. Catalina, A.; Guo, X.; Stefanescu, D.M.; Chuzhoy, L.; Pershing, M.A. Prediction of room temperature microstructure and mechanical properties in lamellar iron castings. *AFS Trans.* **2000**, *94*, 889–912.
9. Urrutia, A.; Celentano, J.D.; Dayalan, R. Modeling and Simulation of the Gray-to-White Transition during Solidification of a Hypereutectic Gray Cast Iron: Application to a Stub-to-Carbon Connection used in Smelting Processes. *Metals* **2017**, *7*, 549. [CrossRef]
10. Fourlakidis, V.; Diószegi, A. A generic model to predict the ultimate tensile strength in pearlitic lamellar graphite iron. *Mater. Sci. Eng. A* **2014**, *618*, 161–167. [CrossRef]
11. Fourlakidis, V.; Diaconu, L.; Diószegi, A. Strength prediction of Lamellar Graphite Iron: From Griffith's to Hall-Petch modified equation. *Mater. Sci. Forum* **2018**, *925*, 272–279. [CrossRef]
12. Diószegi, A. On the Microstructure Formation and Mechanical Properties in Grey Cast Iron. In *Linköping Studies in Science and Technology*; Dissertation No. 871; Jönköping: Jönköping, Sweden, 2004; p. 25. ISBN 91-7373-939-1.
13. Leube, B.; Arnberg, L. Modeling gray iron solidification microstructure for prediction of mechanical properties. *Int. J. Cast Metals Res.* **1999**, *11*, 507–514. [CrossRef]
14. Fourlakidis, V.; Diószegi, A. Dynamic Coarsening of Austenite Dendrite in Lamellar Cast Iron Part 2—The influence of carbon composition. *Mater. Sci. Forum* **2014**, *790–791*, 211–216. [CrossRef]
15. Barkhudarov, M.R.; Hirt, C.W. Casting simulation: Mold filling and solidification—Benchmark calculations using FLOW-3D. In Proceedings of the 7th Conference on Modeling of Casting, Welding and Advanced Solidification Processes, London, UK, 10–15 September 1995.
16. Diószegi, A.; Diaconu, L.; Fourlakidis, V. Prediction of volume fraction of primary austenite at solidification of lamellar graphite cast iron using thermal analyses. *J. Therm. Anal. Calorim.* **2016**, *124*, 215–225. [CrossRef]
17. Svidró, P.; Diószegi, A.; Pour, M.S.; Jönsson, P. Investigation of Dendrite Coarsening in Complex Shaped Lamellar Graphite Iron Castings. *Metals* **2017**, *7*, 244. [CrossRef]
18. Diószegi, A.; Hattel, J. An inverse thermal analysis method to study the solidification in cast iron. *Int. J. Cast Met. Res.* **2004**, *17*, 311–318.
19. Hattel, J.H. *Fundamentals of Numerical Modelling of Casting Processes*, 1st ed.; Polyteknisk Forlag: Lyngby, Denmark, 2005.
20. Diószegi, A.; Fourlakidis, V.; Lora, R. Austenite Dendrite Morphology in Lamellar Cast Iron. *Int. J. Cast Met. Res.* **2015**, *28*, 310–317. [CrossRef]

Article

Effect of Solidification Time on Microstructural, Mechanical and Fatigue Properties of Solution Strengthened Ferritic Ductile Iron

Thomas Borsato [1], Paolo Ferro [1,*], Filippo Berto [2] and Carlo Carollo [3]

[1] Department of Engineering and Management, University of Padova, Stradella S. Nicola, 3, I-36100 Vicenza, Italy; thomas.borsato@phd.unipd.it

[2] Department of Engineering Design and Materials, NTNU, Richard Birkelands vei 2b, 7491 Trondheim, Norway; filippo.berto@ntnu.no

[3] VDP Foundry SpA, via Lago di Alleghe 39, 36015 Schio, Italy; carollo@vdp.it

* Correspondence: paolo.ferro@unipd.it; Tel.: +39-0444-998769

Received: 9 November 2018; Accepted: 21 December 2018; Published: 28 December 2018

Abstract: Microstructural, mechanical, and fatigue properties of solution strengthened ferritic ductile iron have been evaluated as functions of different solidification times. Three types of cast samples with increasing thickness have been produced in a green sand automatic molding line. Microstructural analyses have been performed in order to evaluate the graphite nodules parameter and matrix structure. Tensile and fatigue tests have been carried out using specimens taken from specific zones, with increasing solidification time, inside each cast sample. Finally, the fatigue fracture surfaces have been observed using a scanning electron microscope (SEM). The results showed that solidification time has a significant effect on the microstructure and mechanical properties of solution strengthened ferritic ductile iron. In particular, it has been found that with increasing solidification times, the microstructure becomes coarser and the presence of defects increases. Moreover, the lower the cooling rate, the lower the tensile and fatigue properties measured. Since in an overall casting geometry, same thicknesses may be characterized by different microstructures and mechanical properties induced by different solidification times, it is thought that the proposed methodology will be useful in the future to estimate the fatigue strength of cast iron castings through the numerical calculation of the solidification time.

Keywords: silicon solution strengthened ferritic ductile iron; thickness; solidification time; microstructure; mechanical properties; fatigue; thermal analysis

---

1. Introduction

Since 2012, solution strengthened ferritic ductile irons (SSF-DI) have been introduced in the UNI EN 1563 standard [1]. These alloys are characterized by a fully ferritic matrix, reinforced by the addition of a balanced amount of Silicon, which provides a combination of high strength and ductility [2].

Two of the most important microstructural parameters that are widely used for the estimation of the quality of the ductile iron castings are nodule count and nodularity, as described in the ASTM E2567-16a standard [3]. By increasing the solidification time, the number of graphite particles with a spheroidal shape decreases and their dimension increases; consequently, the mechanical properties are affected. In particular, it has been found that by increasing the section thickness, the nodule count decreases, while the ferrite content increases. Under these conditions, tensile strength and hardness decrease, while elongation at failure increases [4–6].

Longer solidification times cause an increased risk of formation of microstructural defects, with detrimental effect on the mechanical properties [7]. The most common defects that can be found in castings are non-metallic inclusions, shrinkage porosities, and undesired segregation or graphite particles that deviate from the spheroidal shape. Among degenerated graphite morphologies, the branched and interconnected particles, known as chunky graphite (CHG), spiky, exploded, and compacted graphite are the most important and detrimental considering their influence on the mechanical properties [8].

All these defects can be only partly avoided through the optimization of the production process. For example, it was found that adjusted post-inoculation could decrease the dimensions of microshrinkage porosities in heavy section castings [9,10].

Several works [11,12] studied the factors affecting the graphite degeneration in thick walled castings, with particular attention to chunky graphite, which is the most frequent. Although CHG has been considered in many studies, no generally accepted theory for its formation has been found yet. What is known is its detrimental effect on the mechanical properties, with particular reference to the ultimate tensile strength and ductility; but CHG seems also to affect the crack propagation stage during fatigue loadings [13–15]. While in the case of spheroidal graphite particles, decohesion between the nodules and the metal matrix happens, the crack propagates easily through the chunky graphite areas.

On the other side, it has been found by many researchers that the crack initiation stage is more affected by microstructural defects such as non-metallic inclusions, microshrinkage porosities or spiky graphite [16–23]. For example, Borsato et al. [18] proposed a new equation for the assessment of the fatigue limit of ductile iron castings basing on the defects dimensions and the static mechanical properties of the analyzed alloy. In their experiments about the fatigue strength of heavy section ductile iron castings, both Foglio et al. [16] and Ferro et al. [19] observed that the crack initiating defect was a porosity in the vicinity of the sample surface.

While a great effort was spent in the past in order to characterize the traditional ductile irons with ferritic and/or pearlitic matrices, limited data is available in technical papers regarding the new generation ductile irons (SSF-DI) [14,17,24]. Furthermore, only the mechanical properties versus thickness correlation is found, the main drawback of which is that equal thicknesses do not mean necessary equal microstructure and thus mechanical properties. This is the reason why an attempt has been made in the present work to correlate the static and fatigue properties of a solution strengthened ferritic ductile iron with the solidification time supposed to be the most important microstructure-influencing parameter.

2. Experimental Procedure

In this paper, a solid solution strengthened ferritic ductile iron with 3.25 wt % Si was investigated. Melt was prepared in medium frequency induction furnace from pig iron, steel scrap, and ductile cast iron returns. The spheroidizing treatments and the preconditioning were performed in a dedicated ladle by adding 1.2% FeSiMg commercial alloy, containing 1 wt % RE, and 0.3% inoculant (75 wt % Si, 1 wt % Ca) using the sandwich method. After the spheroidizing and inoculation process, and just before pouring the iron into the molds, a metal sample was analyzed by optical emission spectrometry to determine the chemical composition. At the same time, in order to complete the alloy characterization, a thermal analysis was carried out. A standard cup for the thermal analysis, containing the same weight percentage of inoculant of the castings was filled and the cooling curve was then measured by using ITACA MeltDeck^{TM} (ProService Tech, Borgoricco (PD), Italy).

The final chemical composition of the material is shown in Table 1.

Table 1. Final chemical composition (wt %).

C	Si	S	P	Mn	Mg	Ce	Ceq
3.31	3.25	0.008	0.025	0.13	0.050	0.0018	4.40

The carbon content was chosen to be 3.3 wt %, in order to maintain a near eutectic composition, with the Equivalent Carbon calculated using the equation: $C_{eq} = C\% + 0.33(Si\% + P\%)$ [25].

With the aim to evaluate the effect of increasing solidification times on microstructural and mechanical properties, three different geometries, with increasing section thickness were produced in a green sand automatic molding line. 15 molds were produced, each of them containing cast samples with geometries taken from the UNI EN 1563:2012 standard [1]. In particular, the round bar-shaped (type b), the Y-shaped type III and the Y-shaped type IV were used, the relevant wall thicknesses of which were 25, 50, and 75 mm respectively.

From the round bar samples, only tensile test specimens were machined, while from the Y-shaped samples fatigue specimens were also obtained. In particular, the tensile test specimens were characterized by a net diameter of 14 mm, while the fatigue specimens had a rectangular net cross section of 10×15 mm^2.

In order to evaluate the influence of increasing solidification time on the mechanical properties, specimens were taken from the three different cast samples. In particular, in the case of round bar shaped samples with a diameter of 25 mm, the tensile specimens were directly machined. Differently, in the case of Y-shaped samples, a block of about $25 \times 25 \times 175$ mm^3 was cut before the final machining. The positions where the specimens were taken from are shown in Figure 1; it can be noted that from each type IV sample, two specimens were obtained.

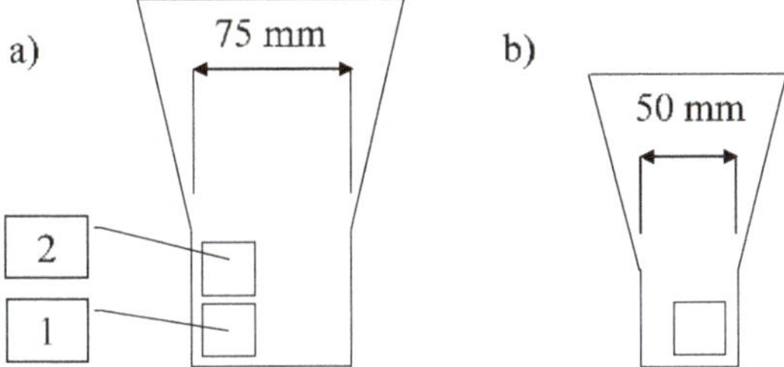

Figure 1. Position of specimens for tensile tests (according to UNI EN 1563:2012) taken from the Y-shaped type IV (**a**) and type III (**b**).

In order to compute the solidification time within each cast sample, numerical analyses were carried out by using the code Novaflow & Solid®(Version 4, Novacast, Ronneby, Sweden). The temperature dependent material properties for ductile iron and green sand have been taken from the database of the numerical code. A size element of 2.5 mm was used for the mesh. The temperature history measured by a virtual thermocouple positioned at the centre of each zone (Figure 1) was used to calculate the solidification time of the whole zone. Tensile tests have been conducted at room temperature according to ISO 6892-1:2016 [26] by using the INSTRON 5500R (Instron, Norwood, MA, USA) tensile test machine under strain rate control. Fatigue tests have been performed according to ASTM E468-18 Standard. A resonant testing machine (RUMUL Testronic 150kN, Russenberger Prüfmaschinen AG, Neuhausen am Rheinfall, Schweiz) was used with a sinusoidal tensile pulsating load at the frequency of about 130 Hz and nominal load ratio R = 0. Tests have been stopped at the total separation of the two parts of the specimens, or after reaching 10^7 cycles. The staircase method was carried out with an applied stress increment of 10 MPa in order to evaluate the fatigue strength corresponding to a fatigue life of 10^7 cycles.

Fatigue tests have been conducted on plain specimens taken from Y-shaped type III and IV cast samples. In particular for each sample, specimens were taken from three levels, numbered consecutively from 1 to 3 going towards the thermal center of the casting, as shown in Figure 2.

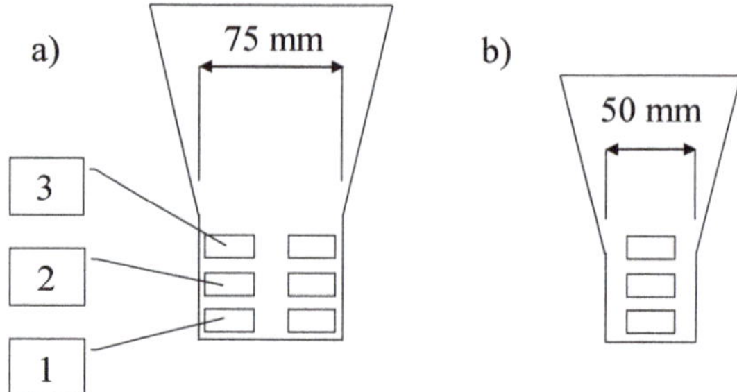

Figure 2. Position of fatigue specimens (rectangular cross section) taken from the Y-shaped type IV (**a**) and type III (**b**).

In the case of type IV sample, six specimens have been obtained, while, due to the smaller dimension, it was possible to take only three specimens from each type III sample. A total of 60 and 30 specimens have been tested for type IV and type III sample, respectively.

Microstructural parameters, such as nodule count, nodule size, nodularity and matrix structure have been evaluated according to the ASTM E2567-16a standard, using an optical microscope and an image analysis software, on polished samples in the unetched and etched (Nital 5%) conditions.

Finally, the fracture surfaces of some broken specimens under fatigue loading have been examined using a scanning electron microscope (Quanta 2580 FEG, FEI, Boston, MA, USA).

3. Results

3.1. Thermal Analysis

The result of thermal analysis is the cooling curve of the alloy from which it is possible to calculate the liquidus temperature, the minimum and maximum eutectic temperature, the temperature at the end of solidification, and finally the temperatures during the solid phase transformation [27,28].

The cooling curve of the SSF-DI obtained from the standard cup and its first time derivative are shown in Figure 3. From a global point of view, two plateaus during the eutectic and the solid-state transformation can be observed.

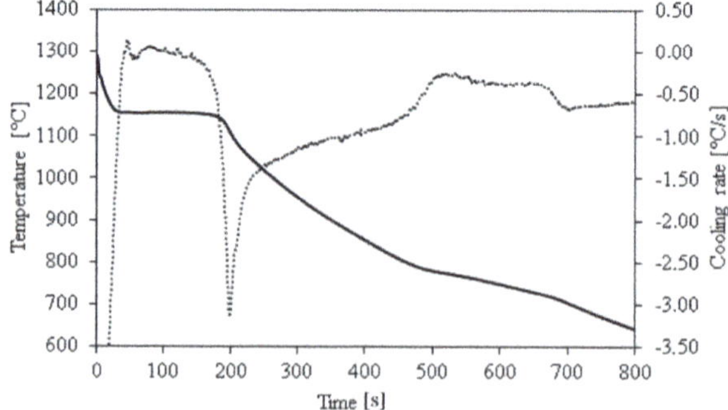

Figure 3. Cooling curve (solid line) and its first derivative (dotted line) of the solution strengthened ferritic ductile iron studied in this work.

From the comparison between the curves of traditional (pearlitic and ferritic-pearlitic) cast irons and solution strengthened ferritic ductile irons during the eutectic and the eutectoid transformation (Figures 4 and 5, respectively), it is visible that there are little variations in the solidification behavior (minimum eutectic temperature about 1153 °C), while the main difference is related to the solid-state transformation.

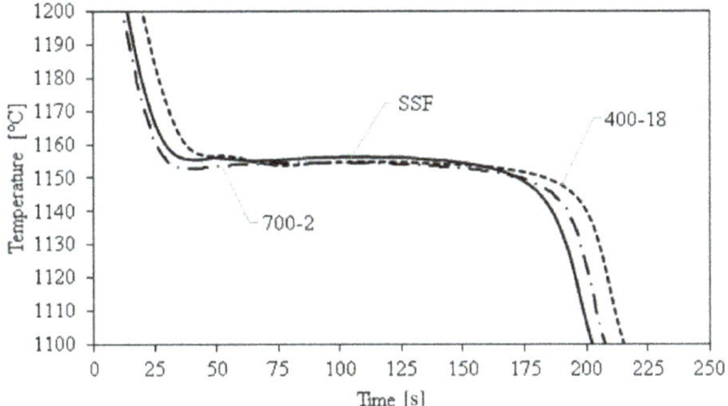

Figure 4. Comparison of temperature profile during the eutectic transformation between traditional and solution strengthened ferritic ductile iron.

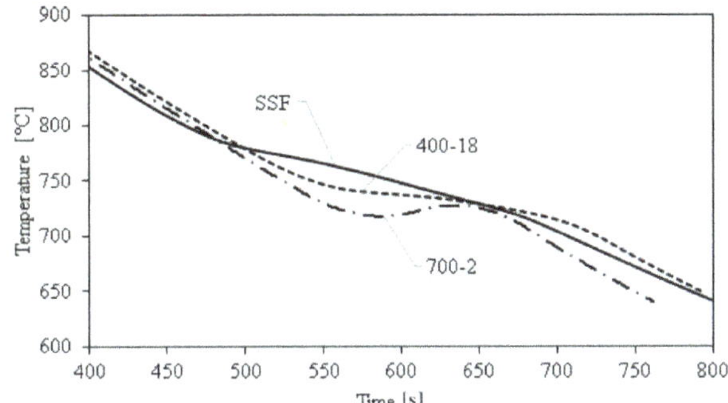

Figure 5. Comparison of temperature profile during the solid-state transformation between traditional and solution strengthened ferritic ductile iron.

As reported in literature [29] the different chemical compositions of the alloys promote the formation of pearlitic and/or ferritic matrix, with differences on the eutectoid temperature and on the cooling rate. In particular, the cooling curve of the GJS 700-2 showed an arrest and a recalescence that is associated to the formation of pro-eutectoid cementite present in pearlite, while in the case of GJS 400-18, the ferritic matrix leads to a higher eutectoid temperature.

Finally, due to the higher silicon content, the eutectoid temperature of the solution strengthened ductile iron is increased with respect to the traditional ferritic grade [30].

3.2. Tensile Tests

Five specimens for each of the four conditions have been tested. Table 2 summarizes the mean values of the results obtained as a function of section thickness and corresponding solidification time, obtained from the numerical analysis.

Compared to other methods used in literature [31], it is important to highlight the fact that, through the solid solution strengthening made by Silicon, it is possible to reach both high strength and ductility. In particular, the ultimate tensile strength is similar to the value of traditional GJS 500-7, while the elongation at failure is close to that of GJS 400-18 cast iron [1].

As expected, by increasing the solidification time, the mechanical properties decrease. This is visible not only between specimens taken from the three different kinds of cast samples, but also from different positions within the same sample. As a matter of fact, the longer the solidification time, the higher the nodules count and nodularity (as described in the paragraph 3.4) and the higher the mechanical properties of the alloy. It means that the solidification time, can be considered as a promising parameter useful to estimate the mechanical properties of the alloy, provided that the constitutive relation 'solidification time versus mechanical property' is experimentally determined.

Table 2. Tensile test results (mean values and standard deviation (in brackets)) of specimens taken from cast samples as a function of section thickness and corresponding solidification time.

Cast Sample	Thickness [mm]	Solidification Time [min]	σ_{UTS} [MPa]	$\sigma_{y\,0.2\%}$ [MPa]	ε_R %
Round bar shaped type b	25	2.5	507 (0.5)	395 (0.5)	19.8 (0.1)
Y-shaped type III	50	9.8	492 (1.0)	389 (0.6)	17.1 (0.1)
Y-shaped type IV (1)	75	16.2	487 (0.6)	386 (0.6)	17.2 (0.9)
Y-shaped type IV (2)	75	22.1	468 (2.1)	375 (1.2)	13.0 (1.0)

3.3. Fatigue Tests

The results of the fatigue tests have been statistically analyzed by using a method that estimates the full S-N curve by considering both the finite life and the run out specimens according to ISO 12107:2012 standard [32]. It is assumed that the finite fatigue life regime consists of an inclined straight line in a logarithm scale with data following a log-normal distribution, while the fatigue endurance region is represented by a horizontal line.

In order to have a graphical comparison of the fatigue behavior, the specimens taken from the different zones within the cast samples are represented with different symbols in Figure 6. The lines represent the 50% fatigue survival probability estimations.

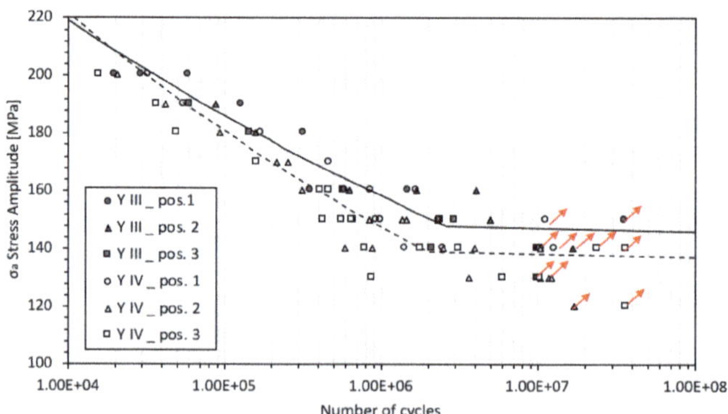

Figure 6. Fatigue life of specimens taken from different position within the cast samples. Solid line and dotted line represent the estimated fatigue curves at 50% survival probability for Y-shaped type III and type IV cast samples respectively. Run out specimens marked with an arrow.

Firstly, the statistical analysis of fatigue data has been performed considering all together the specimens taken from the same cast sample geometry.

In particular it can be observed in Table 3 that, by increasing the section thickness of the casting (going from 50 to 75 mm) the fatigue stress amplitude (σ_a) corresponding to 50% survival probability is lowered by about 7 MPa while the scatter index T_σ, defined as the ratio between the stress amplitude at 10% and 90% survival probability, increases.

Moreover, it can be observed from the graph in Figure 6 that the specimens taken from the position number 1 of the type IV cast samples behave in a similar manner to those obtained from the Y III samples.

Table 3. Fatigue strength at 50% survival probability and scatter index at 10^7 cycles considering cast samples thickness.

Cast Sample	Thickness [mm]	σ_a [MPa]	T_σ
Y-shaped type III	50	145	1.13
Y-shaped type IV	75	138	1.30

In order to better understand the fatigue behavior of the castings, data have been re-analyzed considering no more the thickness, but the solidification time. In particular, three time ranges have been defined based on the numerical simulation results shown in Table 4. Under these conditions, samples taken from position number 3 of the type III sample and from position 1 of the type IV sample, which fell within the same solidification time range, have been analyzed together.

Table 4. Solidification times in the positions shown in Figure 5 obtained from the numerical analysis. The value read at the node in the center of each sample was considered.

Position	Y-Shaped Type III	Y-Shaped Type IV
1	8.5 min	14.5 min
2	11.3 min	18.1 min
3	14.5 min	22.1 min

The obtained fatigue curves at 50% survival probability (the probability that the sample will break at lower load amplitude values) are shown in Figure 7, while the estimated fatigue strength and the scatter index are reported in Table 5 for each range.

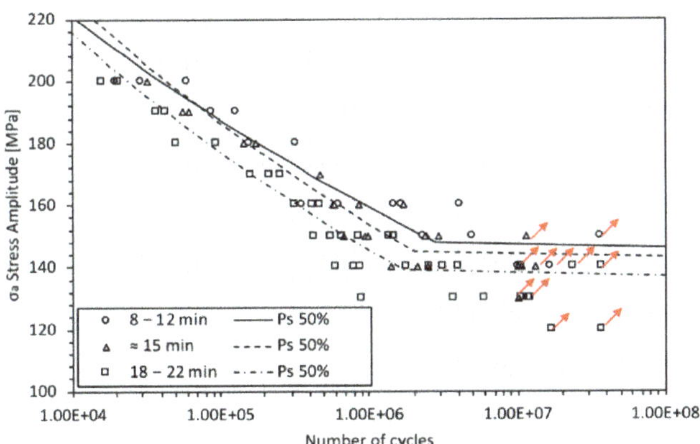

Figure 7. Fatigue life of specimens as a function of solidification time. Lines represent the estimated 50% survival probability curve in the defined solidification time ranges. Run out specimens marked with an arrow.

Table 5. Fatigue strength at 50% survival probability and scatter index at 10^7 cycles considering solidification time range.

Solidification Time	σ_a [MPa]	T_σ
8–12 min	147	1.10
≈15 min	143	1.23
18–22 min	136	1.25

It is important to note that, taking into account the solidification times, it is possible to obtain a more accurate fatigue life estimation in the different positions within the castings, with reduced scatter index compared to the results achieved by considering only the section thickness. This is due to the fact that the statistical analyses were carried out by taking into account specimens with similar cooling conditions, which lead to reduced variability in the mechanical properties.

The relationship between positions within the cast sample and the calculated solidification time is shown in Table 3.

3.4. Microstructure and Fractography

A summary of the microstructural properties is presented in Table 6. In Figure 8, the microstructure of the round bar shaped cast sample is shown, where it can be noted as a fully ferritic matrix with high number of nodules. It can be observed that the solidification time influences the microstructure of the alloys; in particular, it was confirmed that decreasing the cooling rate, fewer nodules are formed with gradually increasing dimensions and lower nodularity. Specimens taken from different cast samples but with the same cooling conditions are characterized by similar microstructural parameters, such as nodule count and nodule diameter.

Table 6. Microstructural properties of samples. Mean values and standard deviation (in brackets).

Cast Sample	Position	Nodule Count [Nodules/mm^2]	Mean Nodule Diameter [μm]	Nodularity
Round bar	-	304	19 (5.5)	93%
Y III	1	115	29 (9)	85%
	2	98	30 (11)	80%
	3	90	28 (10)	77%
Y IV	1	95	29 (9.5)	82%
	2	84	31 (11)	76%
	3	63	33 (13)	70%

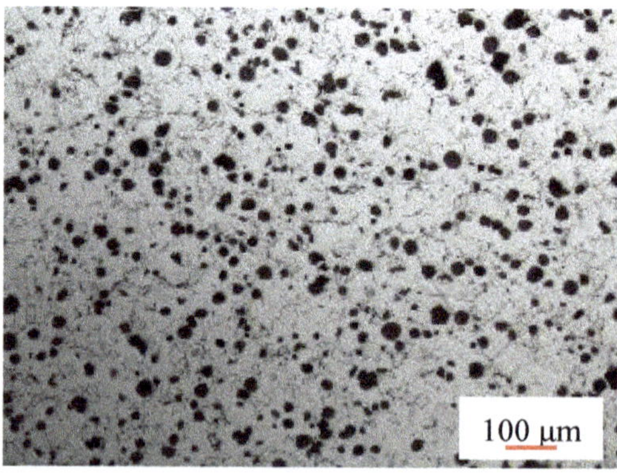

Figure 8. Micrograph of specimen taken from round bar shaped cast sample, etched with Nital 5%.

The increasing in the solidification time could promote also the formation of microstructural defects such as microshrinkage porosities, segregation, or degenerated graphite particles. In fact, going toward the center of the cast samples, it has been found an increasing amount of areas containing branched and interconnected graphite particles, classified as chunky graphite (Figure 9).

Moreover, in the longer to solidify zones, microshrinkage cavities and small area of pearlite, due to segregation of carbide promoter or pearlitizing elements, have been found.

The microstructural properties confirmed what observed during the mechanical tests.

Increasing the section thickness, the solidification time increases and the undercooling decreases. Under these conditions, the number of graphite nodules is lowered, while the risk of forming microstructural defects (microshrinkage porosities, segregations, graphite degenerations, etc.) increases.

The lower nodule count and the presence of such defects could be related to the lower mechanical properties and the higher fatigue scatter index of specimens taken from longer to solidify zones.

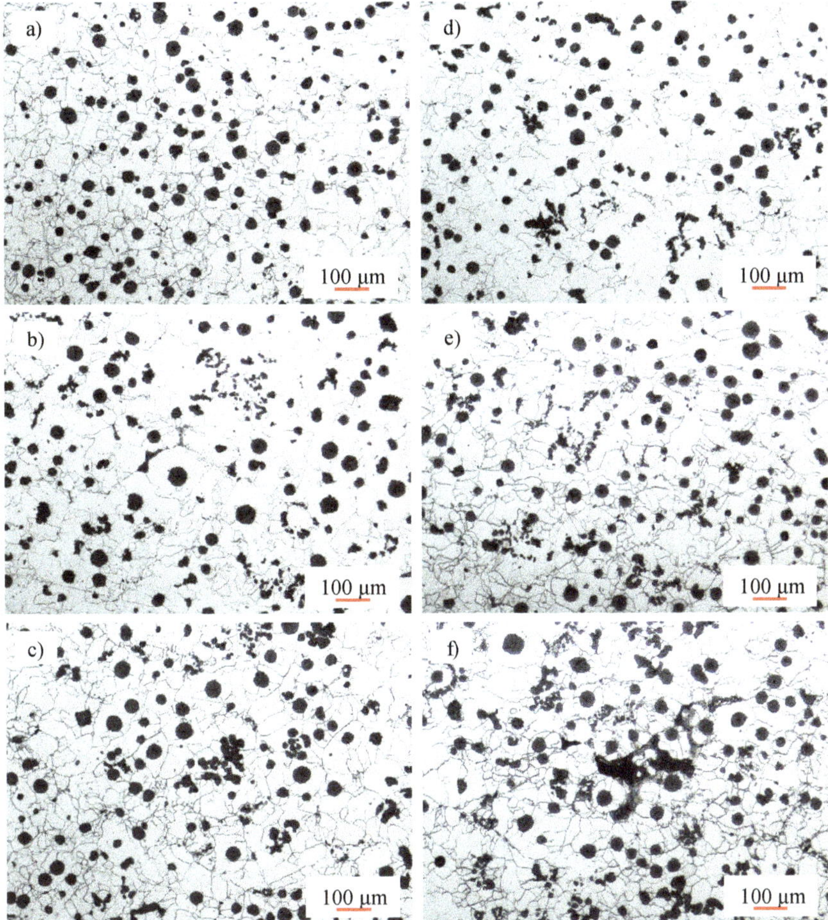

Figure 9. Micrographs of specimens taken from type III cast sample, position 1 (**a**), 2 (**b**) and 3 (**c**) and type IV cast sample, position 1 (**d**), 2 (**e**) and 3 (**f**).

The analyses of fracture surfaces of some fatigue broken specimens revealed that the failure initiated at microshrinkage porosities located near the free surface of the specimens (Figure 10a,b).

The dimensions of such initiating defects ($\sqrt{area}$ parameter) increase with the solidification time, passing from less than 100 μm to more than 400 μm.

It was also found that the failure seems to happen in different ways. Most of the fracture surfaces showed a ductile dimple fracture with microvoids coalescence (Figure 10c,d), however, some areas with brittle transgranular cleavage and intergranular fracture have been even detected (Figure 10e).

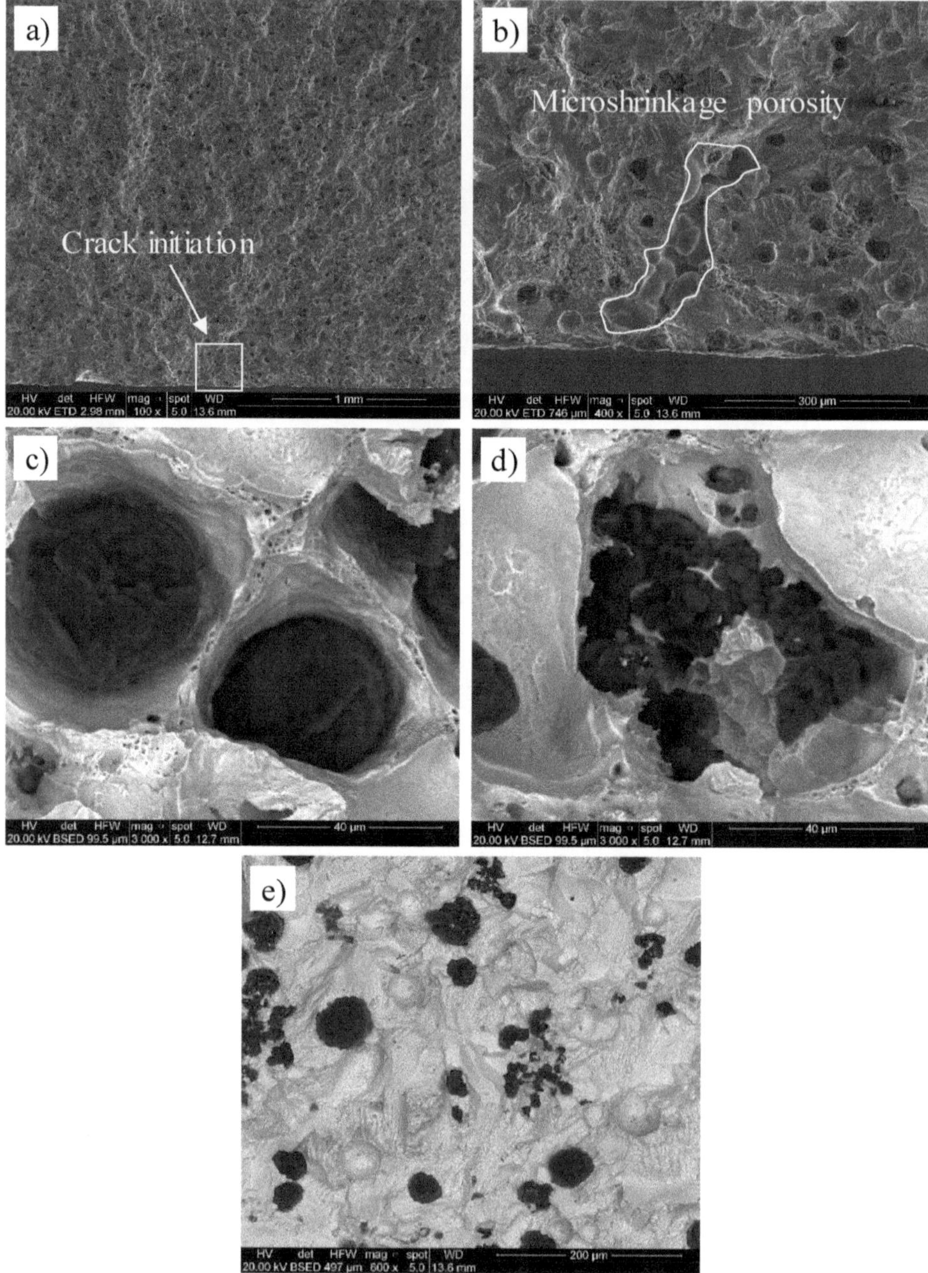

Figure 10. Scanning electron microscope (SEM) images of fracture surfaces showing a panoramic overview of the crack initiation and propagation zone (**a**) and a particular of the crack initiating defect (microshrinkage porosity) (**b**); dimple fracture with microvoids coalescence in the presence of spheroidal graphite nodules (**c**) and degenerated graphite particles (**d**); image of the coexistence of brittle transgranular cleavage and intergranular fracture (**e**).

4. Conclusions

In this paper, the microstructural, mechanical and fatigue properties of a solution strengthened ferritic ductile cast iron characterized by different solidification times have been investigated. It has been confirmed that low cooling rates affect the strength and the ductility of the castings. Due to the high amount of silicon, the microstructure exhibits a fully ferritic matrix, with only small areas of pearlite, due to the segregation of undesired elements, in the thicker sections. Moving toward the zone that takes longer to solidify, the number of graphite nodules decreases and an increasing amount of degenerated chunky graphite is found. The fatigue behavior has been evaluated using specimens taken from different zones inside the castings within defined solidification time ranges. It was confirmed that a better estimation of the fatigue life is achievable by considering the solidification times rather than the section thickness. A decrease of the fatigue strength and an increase of the scatter index were observed with increasing the solidification times, due to the increase in the defects dimensions.

All the samples showed shrinkage porosities, which have been identified as crack initiation sites. The fracture surface revealed that the crack propagates easier through the areas with chunky graphite compared to areas with spheroidal graphite nodules. Most of the fracture surface showed a ductile dimple fracture with microvoids coalescence, but some areas with brittle transgranular cleavage and intergranular fracture were also detected. Finally, it was observed that the solidification time can be really considered as a microstructure-influencing parameter. Basing on the intrinsic relation between microstructure and mechanical properties, it will be possible in the future to estimate the static and fatigue strength of the alloy simply by a numerical computation of the solidification time, provided that the constitutive relation 'solidification time versus mechanical property' is 'a-priori' determined.

Author Contributions: Conceptualization, P.F., T.B. and F.B.; methodology, P.F. and T.B.; validation, T.B.; formal analysis, T.B.; investigation, T.B. and C.C.; F.B.; writing—original draft preparation, T.B.; writing—review and editing, P.F. and F.B.; visualization, P.F., T.B. and F.B.; supervision, P.F. and F.B.

Funding: This research received no external funding.

Acknowledgments: The authors would like to thank VDP Fonderia SpA for the material supply and technical support of the project.

Conflicts of Interest: The authors declare no conflict of interest.

References

1. UNI EN 1563:2012. *Founding—Spheroidal Graphite Cast Irons*; UNI: Milano, Italy, 2012.
2. Larker, R. Solution strengthened ferritic ductile iron ISO 1083/JS/500-10 provides superior consistent properties in hydraulic rotators. *Overseas Foundry* **2009**, *6*, 343–351.
3. ASTM E2567-16a. *Standard Test Method for Determining Nodularity and Nodule Count in Ductile Iron Using Image Analysis*; ASTM International: West Conshohocken, PA, USA, 2016.
4. Ceschini, L.; Morri, A.; Morri, A. Effects of Casting Size on Microstructure and Mechanical Properties of Spheroidal and Compacted Graphite Cast Irons: Experimental Results and Comparison with International Standards. *J. Mater. Eng. Perform.* **2017**, *26*, 2583–2592. [CrossRef]
5. Shinde, V.D.; Ravi, B.; Narasimhan, K. Solidification behaviour and mechanical properties of ductile iron castings with varying thickness. *Int. J. Cast Met. Res.* **2012**, *25*, 364–373. [CrossRef]
6. Bockus, S.; Venckunas, A.; Zaldarys, G. Relation between section thickness, microstructure and mechanical properties of ductile iron castings. *Mater. Sci.* **2008**, *14*, 115–118.
7. Minnebo, P.; Nilsson, K.-F.; Blagoeva, D. Tensile, compression and fracture properties of thick-walled ductile cast iron components. *J. Mater. Eng. Perform.* **2007**, *16*, 35–45. [CrossRef]
8. Ecob, C.M. *A Review of Common Metallurgical Defects in Ductile Cast Iron*; Customer Services Manager, Elkem AS, Foundry Products Division: Oslo, Norway, 2005.
9. Regordosa, N. Llorca-Isern, Microscopic Characterization of Different Shrinkage Defects in Ductile Irons and their Relation with Composition and Inoculation Process. *Int. J. Met.* **2017**, *11*, 778–789.
10. Borsato, T.; Ferro, P.; Berto, F.; Carollo, C. Fatigue strength improvement of heavy-section pearlitic ductile iron castings by in-mould inoculation treatment. *Int. J. Fatigue* **2017**, *102*, 221–227. [CrossRef]

11. Källbom, R.; Hamberg, K.; Wessén, M.; Björkegren, L.E. On solidification sequence of ductile iron castings containing chunky graphite. *Mater. Sci. Eng. A* **2005**, *413*, 346–351. [CrossRef]

12. Ferro, P.; Fabrizi, A.; Cervo, R.; Carollo, C. Effect of inoculant containing rare earth metals and bismuth on microstructure and mechanical properties of heavy section near-eutectic ductile iron castings. *J. Mater. Process. Technol.* **2013**, *213*, 1601–1608. [CrossRef]

13. Foglio, E.; Lusuardi, D.; Pola, A.; La Vecchia, G.M.; Gelfi, M. Fatigue design of heavy section ductile irons: Influence of chunky graphite. *Mater. Des.* **2016**, *111*, 353–361. [CrossRef]

14. Borsato, T.; Berto, F.; Ferro, P.; Carollo, C. Influence of solidification defects on the fatigue behaviour of heavy-section silicon solution–strengthened ferritic ductile cast irons. *Fatigue Fract. Eng. Mater. Struct.* **2018**. [CrossRef]

15. Mourujärvi, A.; Widell, K.; Saukkonen, T.; Hänninen, H. Influence of chunky graphite on mechanical and fatigue properties of heavy-section cast iron. *Fatigue Fract. Eng. Mater. Struct.* **2009**, *32*, 379–390.

16. Foglio, E.; Gelfi, M.; Pola, A.; Goffelli, S.; Lusuardi, D. Fatigue Characterization and Optimization of the Production Process of Heavy Section Ductile Iron Castings. *Int. J. Met.* **2017**, *11*, 33–43. [CrossRef]

17. Borsato, T.; Ferro, P.; Berto, F.; Carollo, C. Mechanical and fatigue properties of heavy section solution strengthened ferritic ductile iron castings. *Adv. Eng. Mat.* **2016**, *18*, 2070–2075. [CrossRef]

18. Borsato, T.; Ferro, P.; Berto, F. Novel method for the fatigue strength assessment of heavy sections made by ductile cast iron in presence of solidification defects. *Fatigue Fract. Eng. Mater. Struct.* **2018**. [CrossRef]

19. Ferro, P.; Lazzarin, L.; Berto, F. Fatigue properties of ductile cast iron containing chunky graphite. *Mater. Sci. Eng. A* **2012**, *554*, 122–128. [CrossRef]

20. Collini, L.; Pirondi, A. Fatigue crack growth analysis in porous ductile cast iron microstructure. *Int. J. Fatigue* **2014**, *62*, 258–265. [CrossRef]

21. Bleicher, C.; Kaufmann, H.; Melz, T. Fatigue Assessment of Nodular Cast Iron with Material Imperfections. *SAE Int. J. Engines* **2017**, *10*, 340–349. [CrossRef]

22. Luo, J.; Bowen, P.; Harding, R.A. Evaluation of the fatigue behavior of ductile irons with various matrix microstructures. *Met. Mater. Trans. A* **2002**, *33*, 3719–3730. [CrossRef]

23. Nadot, Y.; Mendez, J.; Ranganathan, N. Influence of casting defects on the fatigue limit of nodular cast iron. *Int. J. Fatigue* **2004**, *26*, 311–319. [CrossRef]

24. Kasvayee, K.A.; Ghassemali, E.; Svensson, I.L.; Olofsson, J.; Jarfors, A.E.W. Characterization and modeling of the mechanical behavior of high silicon ductile iron. *Mater. Sci. Eng. A* **2017**, *708*, 159–170. [CrossRef]

25. Davis, J.R. (Ed.) *ASM Specialty Handbook: Cast Irons*; ASM International: Almere, The Netherlands, 1996.

26. ISO 6892-1:2016. *Metallic Materials—Tensile Testing—Part 1: Method of Test at Room Temperature*; ISO: Geneva, Switzerland, 2016.

27. Stefanescu, D.M. Thermal analysis—Theory and applications in metalcasting. *Int. J. Met.* **2015**, *9*, 7–22. [CrossRef]

28. Shinde, V.D. Thermal Analysis of Ductile Iron Casting. *IntechOpen* **2018**. [CrossRef]

29. Sertucha, J.; Larranaga, P.; Lacaze, J.; Insausti, M. Experimental investigation on the effect of copper upon eutectoid transformation of as-cast and austenitized spheroidal graphite cast iron. *Int. J. Met.* **2010**, *4*, 51–58. [CrossRef]

30. Labrecque, C.; Gagne, M. Review ductile iron: Fifty years of continuous development. *Can. Metall. Q.* **1998**, *37*, 343–378. [CrossRef]

31. Zykova, A.; Lychagin, D.; Chumaevsky, A.; Popova, N.; Kurzina, I. Influence of Ultrafine Particles on Structure, Mechanical Properties, and Strengthening of Ductile Cast Iron. *Metals* **2018**, *8*, 559. [CrossRef]

32. ISO 12107:2012. *Metallic Materials—Fatigue Testing—Statistical Planning and Analysis of Data*; ISO: Geneva, Switzerland, 2012.

Article

Determination of the Effective Elastic Modulus for Nodular Cast Iron Using the Boundary Element Method

Adrián Betancur [1], Carla Anflor [1,*], André Pereira [2,*] and Ricardo Leiderman [3]

[1] Group of Experimental and Computational Mechanics, University of Brasília, 72.405-610 Gama, Brazil; aabetanc@gmail.com

[2] Engineering School, Fluminense Federal University, Rua Passo da Patria 156, 24210-240 Niterói, Brazil

[3] Institute of Computing, Fluminense Federal University, Rua Passo da Pátria 156, 24210-240 Niterói, Brazil; leider@ic.uff.br

* Correspondence: anflor@unb.br (C.A.); andremaues@id.uff.br (A.P.); Tel.: +55-61-3107-8916 (C.A.); +55-21-9-8242-2070 (A.P.)

Received: 21 June 2018; Accepted: 21 July 2018; Published: 15 August 2018

Abstract: In this work, a multiscale homogenization procedure using the boundary element method (BEM) for modeling a two-dimensional (2D) and three-dimensional (3D) multiphase microstructure is presented. A numerical routine is specially written for modeling nodular cast iron (NCI) considering the graphite nodules as cylindrical and real geometries. The BEM is used as a numerical approach for solving the elastic problem of a representative volume element from a mean field model. Numerical models for NCI have generally been developed considering the graphite nodules as voids due to their soft feature. In this sense, three numerical models are developed, and the homogenization procedure is carried out considering the graphite nodules as non-voids. Experimental tensile, hardness, and microhardness tests are performed to determine the mechanical properties of the overall material, matrix, and inclusion nodules, respectively. The nodule sizes, distributions, and chemical compositions are determined by laser scanning microscopy, an X-ray computerized microtomography system (micro-CT), and energy-dispersive X-ray (EDX) spectroscopy, respectively. For the numerical model with real inclusions, the boundary mesh is obtained from micro-CT data. The effective properties obtained by considering the real and synthetic nodules' geometries are compared with those obtained from the experimental work and the existing literature. The final results considering both approaches demonstrate a good agreement.

Keywords: boundary element method (BEM); periodic boundary conditions; representative volume elements (RVEs); effective elastic properties; homogenization

---

1. Introduction

The multiscale homogenization procedure for microstructural material is an important topic in the science and engineering of materials. According to Nguyen et al. [1], computer modeling is the most efficient technique for this purpose, and it has been a subject of interest in recent years. One of the main challenges for multiscale modeling is in developing numerical models that predict the behavior of microstructures with high accuracy and efficiency. Many works on this subject have been developed over the years [2–5]. According to Pundale et al. [6], microstructural materials are considered to be composites for the numerical modeling approach, where the challenge relies on coupling the individual behavior of each constituent to obtain an overall response.

Achenbach and Zhu [7,8] carried out numerical studies with multiphase media using a unit cell model to study the variation of the interface parameters and to estimate the effect of distribution of stresses in both the matrix and in the fibers. Moorthy and Ghosh [9] developed a finite element (FE) model in conjunction with Voronoi cells to examine the small deformation of arbitrary heterogeneous material in a 2D model. The influences of the shape, size, orientation, and distribution of inclusions on the micro- and macroscopic responses were also investigated. The results indicated that the orientation of every grain did not play as important a role as the size of the inclusions in the response of the overall properties.

Effective properties of macrostructural materials using synthetic geometries (i.e., regular geometries such as cylinders, spheres, and ellipsoids) of different sizes and different mechanical properties were studied by Yao et al. [10], who used a 2D formulation in the boundary element method (BEM) to demonstrate that the BEM formulation may be more suitable for interface analysis when compared to the FE method. Using the following two approaches, Zheng et al. [11] studied a solid whose macrostructure comprised fluid-filled pores: the superposition method and a multi-subdomain method 2D BEM model using subregions. It was shown that the multi-subdomain method (with subregions) was more efficient and accurate for determining the effective properties. Gitman et al. [12] discussed the periodicity of the material, and they established the no-wall-effect concept, which refers to the inability of inclusions to penetrate through the sample borders within a representative volume element (RVE). According to Gitman [12], the periodicity of a material considers that a material experiencing no-wall effects due to the RVE is a representative volume and should, therefore, represent any part of the material. Despite the many experimental and computational methods devoted to the study of the effective elastic modules of nodular cast iron (NCI) [6,13–15], it is important to highlight that computational methods are much more attractive from the point of view of saving time and reducing the operational costs.

Numerical homogenization was used based on the concept of the RVE for determining the effective properties. Through this method, the statistical data of many RVEs can be obtained, and by averaging the values of each distribution of spherical inclusions, it is possible to find the global response of the material [16]. According to Zohdi [16], this technique is more trustworthy than performing individual direct simulations. Buroni and Marczak [13] developed a homogenization procedure using a 2D BEM as a numerical approach to determine the effective modulus of NCI. In this work, the inclusions were considered to be voids, which were discretized with a single special hole element. This new numerical approach demonstrated good accuracy and low computational cost. Carazo et al. [17] performed an RVE study to determine the effective properties of NCI. The authors imposed rectangular and hexagonal geometric shapes of the RVE; the numerical results were compared to analytical expressions, and the authors concluded that the graphite fraction has the largest influence on the effective modulus. They realized that the effective properties are independent of the RVE shape. Fernandino et al. [15] recently used the FE method and a multiscale analysis to determine the effective elastic proprieties of NCI where a complete experimental characterization was introduced and used to evaluate the numerical results.

In this paper, the main objective is to develop a homogenization process for NCI. Nodular cast iron, also known as spheroidal graphite cast iron (SGI), was chosen to develop a numerical model of homogenization. The main reason is the morphology of the microstructure, which contains nearly spherical inclusions of graphite embedded in a homogeneous ferritic matrix. The nodular graphite has a high carbon content, while the matrix medium has a ferritic, pearlitic, ferritic-pearlitic, or austenitic structure [18]. For the numerical homogenization process, two approaches were employed: the first one models the nodules as a synthetic geometry, and the second one models them as their real shapes obtained via microtomography. A numerical routine using the BEM for the homogenization analysis was supported by experimental studies, such as microtomography by X-rays (micro-CT), tensile tests, hardness tests, microhardness tests, micrographic analyses, and scanning electron microscopy equipped with energy-dispersive X-ray (SEM-EDX) analyses. The main goal of this investigation is to determine the influence on the effective properties obtained when considering the nodules as synthetic and real geometries in the homogenization process.

2. Numerical Model

A multiscale numerical routine was written based on the BEM as the numerical approach for determining the effective Young's modulus of heterogeneous materials with linear elastic behavior. The graphite was modeled as a second constituent by using the subregion methodology provided by BEM. The subregion formulation takes into account the stiffness of nodular inclusions within the ferritic homogeneous matrix of NCI. To accelerate the convergence of the homogenization process, periodic boundary conditions were also implemented in the numerical routine, particularized to plane stress (PT).

2.1. The BEM and Subregions

The BEM is essentially based on the domain's boundary, which is partitioned in a coarse mesh where the differential governing equation is numerically integrated [19,20]. This numerical approach has all the unknown variables only on the boundary, meaning that no information from inside the domain is required. Figure 1 illustrates a boundary value problem considering both an RVE by sub-region and a no-wall-effect concept [12]. The ferritic matrix is denoted by the subscript 1, and the graphitic nodules are indicated by the subscript 2.

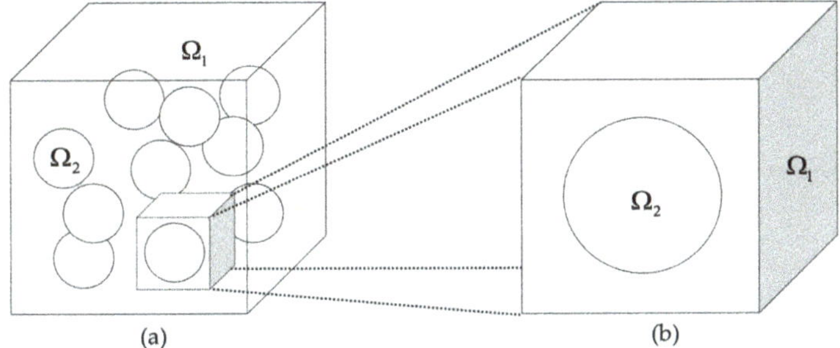

Figure 1. BEM using subregions (or multi-domains): (**a**) RVE for NCI and (**b**) RVE for a unitary cell.

The governing equation for a linear elastic material is presented in Equation (1):

$$u(x) = \int_{\Gamma_i} U_{ij}(x,y)t(y)t(y)ds_y - \int_{\Gamma_i} T_{ij}(x,y)u(y)ds_y \tag{1}$$

where U and T are the components of displacement and tractions containing the fundamental solutions for the nodal displacement (u) and tractions (t), respectively. The fundamental solutions for a 2D elastic model are as follows:

$$U_{ij}(x,y) = -\frac{1}{8\pi\mu}\left[(3-4v)\log\frac{1}{r}\delta_{ij} + r_i r_j\right] \tag{2}$$

$$T_{ij}(x,y) = -\frac{1}{4\pi(1-v)r}\left\{\frac{\partial r}{\partial n}\left[(1-2v)\delta_{ij} + 2r_i r_j\right] + (1-2v)\left(n_i r_j - n_j r_i\right)\right\} \tag{3}$$

where μ and v are the shear stress and Poisson's ratio, respectively; r is the distance between the source point and a field point; and δ stands for the Kronecker delta. Once the external boundary conditions are imposed, the system is integrated over the total number of nodes (T_n), resulting in a linear system of equations:

$$\sum_{i=1}^{T_n}\left[H_{ij}u_i\right] = \sum_{i=1}^{T_n}\left[G_{ij}t_i\right] \rightarrow [H](u) = [G](t) \tag{4}$$

where H and G are the arrays of displacement and tractions, respectively. Every subregion is discretized by discontinuous quadratic elements. Therefore, the BEM equations for each subregion were implemented according to Brebbia and Dominguez [19] and Aliabadi [20].

For the homogenization process to succeed, the influence of the boundary conditions imposed on the problem must be taken into account in order to increase the accuracy and convergence. In this sense, three classical methods may be used to impose the boundary conditions [21]: (1) a linear displacement boundary condition (Dirichlet condition); (2) a constant traction boundary condition (Neumann condition); and (3) a periodic boundary condition (PBC). Some works [21–23] emphasize the importance of imposing the PBC due to its efficiency and accuracy for the final results of analyses in microstructural material. In this work, a PBC that considers the Dirichlet condition was implemented to ensure good accuracy in the homogenization procedure.

2.2. Constitutive Law

The constitutive linear equation of an elastic material is written in index notation as:

$$\sigma_{ij} = C_{ijkm}\varepsilon_{km} \tag{5}$$

where C_{ijkm} represents the components of the fourth-order constitutive tensor, which relates the stress and strain tensors, and it is related to the mechanical properties according to Equation (6):

$$C_{ijkm} = \frac{E}{1 - v^2}\left[(1 - v)\mathrm{I}_{ijkm} + v\left(\mathrm{I}_{ij} \times \mathrm{I}_{ij}\right)\right] \tag{6}$$

where E and v denote the Young's modulus and Poisson's ratio, respectively. In our problem using subregions, these parameters are given by:

$$E = \begin{cases} E_{matriz} & and & v_{matriz} & if & (x,y) \in \Omega_{matrix} \\ E_{nodule} & and & v_{nodule} & if & (x,y) \in \Omega_{nodule} \end{cases} \tag{7}$$

where E_{matrix} and v_{matrix} are the Young's modulus and Poisson's ratio, respectively, for the ferritic matrix, while E_{nodule} and v_{nodule} are the Young's modulus and Poisson's ratio for the graphite nodule.

2.3. The Plane Stress (PT) and Plane Strain (PS) Hypotheses and the Mean Field Theory

The PT and PS hypotheses are two considerations that should be taken into account to decrease the computational costs as well as to minimize the number of variables for a 2D analysis. The macroscopic properties of a heterogeneous material are typically determined using Equation (8), where the micro- and macro-couplings are implicitly defined. Equations (9) and (10) establish the relationship between the infinitesimal macroscopic stress and strain tensors and the averages of the stress and strain in the RVE [24]:

$$\langle\sigma_{ij}\rangle_\Gamma = C^*_{ijkm}\langle\varepsilon_{km}\rangle_\Gamma \tag{8}$$

$$\varepsilon_{ij}^M = \langle\varepsilon_{ij}\rangle_\Gamma \tag{9}$$

$$\sigma_{ij}^M = \langle\sigma_{ij}\rangle_\Gamma \tag{10}$$

where $\langle\rangle$ stands for the average stress taken over the boundary Γ. It is the spatial average operator, which is defined as follows:

$$\langle\bullet\rangle = \frac{1}{|\Gamma|}\int_\Gamma \bullet\, d\Gamma \tag{11}$$

when $\bullet$ is the mechanical property to be evaluated. According to Zohdi et al. [25], a specimen can be computationally tested to permit the solution to the elastic problem for determining the stress and strain micro fields. Solving the system of equations shown in Equation (8) determines the constants for the matrix of elasticity. The microstructural sample material is subjected to the same boundary conditions, resulting in a field of tensions and deformations that are distributed uniformly as an entire homogeneous body [13,24].

3. Methodology and Materials

In this work, the graphite nodules were modeled numerically using two different procedures. The first one employs a statistical analysis, assuming that nodular graphite has a synthetic geometry (circular shape), and the second procedure assumes that the nodular graphite has the real geometry obtained by X-ray microtomography. The morphological characterization and mechanical properties were determined through experimental tests, which were used to support and validate the proposed numerical homogenization.

3.1. X-Ray Microtomography

To validate the proposed methodology, digital images were acquired using micro-CT. The samples were scanned using a ZEISS X radial 510 Versa, high-resolution, micro-CT scanner system (Versa 510, Bregnerødvej, Birkerød, Denmark). This system generates X-rays with a cone-beam geometry, and a 3D digital object can be visualized from the acquired radiograms. The small samples of NCI have dimensions of 6 mm × 7 mm × 24 mm. Each sample was then mounted on a rotating cylindrical support inside the X-ray chamber and positioned so that the X-ray lines were perpendicular to their axis of rotation. Then, the scanning parameters were adjusted and scanned for 15 s. The geometry obtained by micro-CT was straightforwardly exported to generate a BEM mesh, making it possible to use the real geometry of the microstructure.

Through the micro-CT images, it was also possible to determine the shape coefficient K_f ($K_f = D_1/D_2$) and the characteristic parameter r/d ($r = (D_1 + D_2)/4$). Here, D_1 and D_2 are the smallest and the largest diameter, respectively, measured inside the graphite nodule; r is the average radio; and d is the average distance measured between two graphite nodules [26,27].

To obtain an approximate representation of the characteristic parameters of the NCI microstructure, a statistical analysis was implemented by random selection of graphite nodules over the analysis zone. All data were tabulated in histograms.

3.2. Microstructure of NCI

A micrographic analysis was used to characterize the spatial distribution of graphite nodules and the average nodule size. Micrograms were obtained from NCI blocks with dimensions of 5 mm × 5 mm × 5 mm. The surface was polished and etched with nital at 3%, and the samples were then mounted on a sample holder and analyzed in a vacuum according to ASTM E112-96 standard. Three micrograms were obtained with scales of 400, 100, and 20 μm, and the resulting images were used to measure the amount, area, and overall size of the graphite nodules.

3.3. Hardness Tests

Vickers and Brinell hardness testing was performed to measure the hardness of both the ferritic matrix and the graphite nodules in line with ASTM E92-16 standard. By using the micro-indentation methodology [28], it was possible to determine the Young's modulus for both the matrix and the graphite nodules. In this work, the measurements were carried out with an FM 700 Vickers microhardness apparatus (FM 700 Vickers, Karola Miarki 12, Gliwice, Polska), equipped with a glass camera with a 100× magnification capability and which had a load of 10 g. The average hardness of the ferritic bulk was determined using a Brinell hardness (HB) test, in line with ASTM E92-16, with a semiautomatic HB tester (ZHU250 HB, Kennesaw, GA, USA). Afterwards, the tensile strength of NCI was determined through the standard correlations between the HB hardness and tensile strength. Figure 2a,b depict the shallow indentations made in the microhardness tests, revealing that the graphite nodule is softer than the matrix due to the difference in highlights between the areas of the hardness impression. The HB test is presented in Figure 2c. In this case, the spherical indentations are larger than the Vickers hardness HV indention, meaning that the hardness value only indicates the toughness of the bulk material.

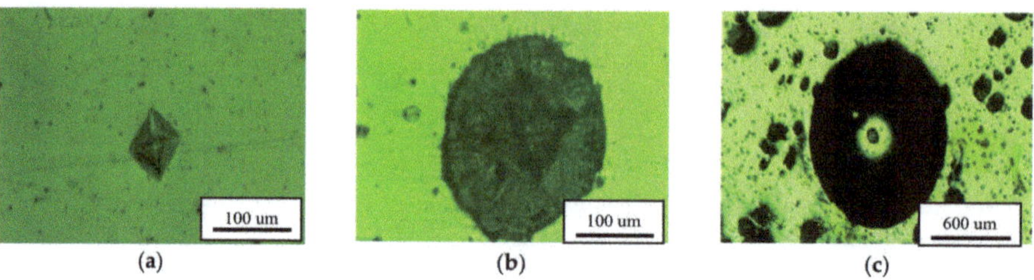

(a)   (b)   (c)

Figure 2. Hardness tests: (a) HV matrix indentation; (b) HV graphite nodular indentation; and (c) HB indentation.

The mathematical model discussed in [28] consists of identifying the geometric parameters of the indented surface. Then, from the geometrical parameters of the hardness impressions and the mechanical and geometrical properties of the indenter, it is possible to determine the Young's modulus according to Equation (12):

$$E = \frac{1 - v^2}{\frac{1}{E_r} - \frac{1 - v_i^2}{E_i}}$$ (12)

where v_i is the Poisson's ratio of the indenter, v is the Poisson's ratio of the sample, E_i is the Young' modulus of the indenter, E_r is the reduced Young's modulus of the sample, and E is the Young's modulus of the sample. The parameters used for estimating the effective Young's modulus are listed in Table 1.

Table 1. Parameters used for estimating the experimental Young's modulus.

Parameter	Value
v	0.30
v_i	0.07
E_i	1411 GPa
Load applied HV	10 gf

The values of E_r were extracted by using Equation (13):

$$E_r = \frac{\omega P_{max}\sqrt{\pi}}{2h_s\sqrt{A}}$$ (13)

where ω is a constant that depends on the geometry of the indenter; P_{max} is the maximum load; h_s is the amount of sink-in, which is assumed to be approximately equal to the final depth h_f; and A is the projected area of the elastic contact. Due to the uncertainty in the Vickers system and the structurally inhomogeneous sample, a statistical test was developed using micro-indentation patterns containing 50 indentations over the analysis zone.

3.4. Tensile Tests

Tensile testing was performed on sub-size test samples, which were obtained by machining the NCI blocks according to the specifications in ASTM E8/E8M-09 standard, using an Instron 8801 test machine (Instron, Glenview, IL, USA) [±100 kN (22,500 lbf)] with a 2620-601 dynamic gauge extensometer (Instron, Glenview, IL, USA). The Young's modulus, tensile strength, and ultimate tensile strength were determined from the experimental stress-strain curve.

3.5. Scanning Electron Microscope Equipped with EDX

The quantitative compositional information for NCI was determined by using a scanning electron microscope system equipped with energy dispersive X-ray analyzer system (JEOL JSM 6610 SEM-EDX, Musashino, Tokyo, Japan). The samples were polished, treated, and placed in a vacuum chamber for visualization of the surface. The samples were analyzed at 100×, and the microstructural examinations were carried out for a representative sample with a large number of graphite inclusions.

3.6. Computational Homogenization

The homogenization concept is based on a multiscale analysis that establishes the relationships between macro- and micro-scale entities [29]. The micromechanical analysis typically focuses on the concepts of RVEs; the analysis involves an average representative volume extracted from the micro-heterogeneity in an infinitesimal-volume sample [1,12,24]. This sample contains a reasonable representation of the microstructure's material; therefore, the material properties are obtained through average volumes of the respective micro fields obtained from an RVE [12,24]. The numerical homogenization was developed following two procedures.

3.6.1. Case I: Graphite Nodules Considered as Synthetic Geometry

The synthetic homogenization procedure was developed using the characteristic parameters obtained from statistical analyses based on morphological characterizations. There is no definitive way in which to establish an RVE [16].; the most frequently used procedure currently involves either fixing the nominal RVE size and then increasing the number of graphite nodules (Figure 3) or increasing the size of the RVE so that the number of heterogeneities changes by a fixed percentage.

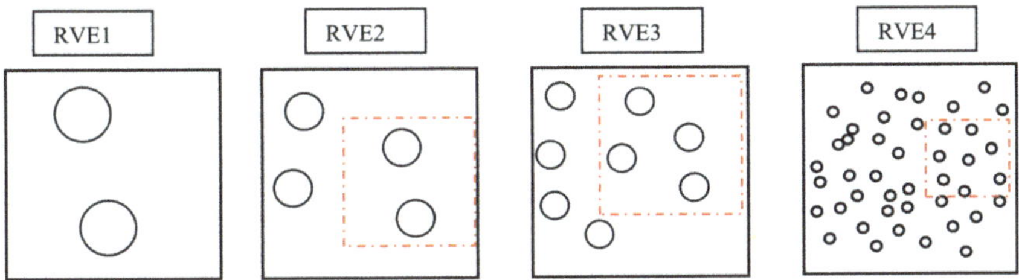

Figure 3. The RVE with different sizes of graphite nodules.

In Figure 3, RVE1 ⊂ RVE2 ⊂ RVE3 ⊂ RVE4. The discretization strategy of the microstructure was assumed to be a square sheet for the 2D RVE case and a cube for the 3D RVE case. In both cases, a constants dimension, containing a specific number of nodules, was defined. The characteristic parameter obtained with the procedure described in Section 3.1 was used to model NCI. The nodules/matrix ratio of the sample was obtained by morphological characterization using image analysis during the binarization process, which establishes the percentage of each phase. The homogenization model employed samples with the following numbers of nodules: 4, 6, 8, 10, 15, 20, 25, 30, 40, 45, 50, 55, and 60.

3.6.2. Case II: Graphite Nodules Considered as Real Geometry

Figure 4 presents the homogenization procedure used to estimate the effective Young's modulus considering a real morphology. This procedure involves the generation of several RVEs with different sizes and in random coordinates over the sample (Figure 4a) [12]. The RVE sizes were obtained from numerical tests, where constant displacements were imposed as a boundary condition. Furthermore, the micrograms obtained by micro-CT were used to obtain the forms and distribution of the nodules (Figure 4b). In the sequence, a 2D mesh that represents the inclusions of the irregular geometry inside the ferritic matrix was generated (Figure 4c).

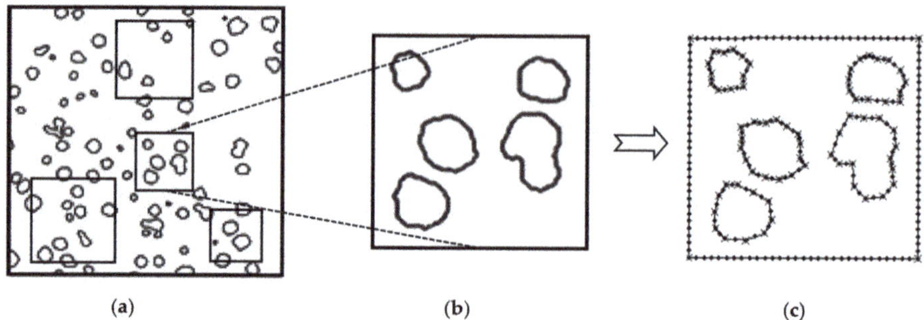

(a)   (b)   (c)

Figure 4. The RVE for NCI: (**a**) sample, (**b**) random RVE, and (**c**) 2D mesh.

The boundary was discretized with a discontinuous quadratic element. Moreover, the dimensions of the RVE were progressively increased until the effective Young's modulus was reached.

4. Results and Discussion

The results obtained from the experimental and numerical tests described in the previous section are presented below.

4.1. X-Ray Microtomography

According to the micro-CT images, the graphite nodules exhibited a shape that is approximately a circle. In this sense, the shape parameter (K_f) and the characteristic parameter (r/d) were estimated through statistical testing. Figure 5 illustrates the nodule distributions in an NCI sample. The results for K_f and the r/d ratio are presented in a histogram graph.

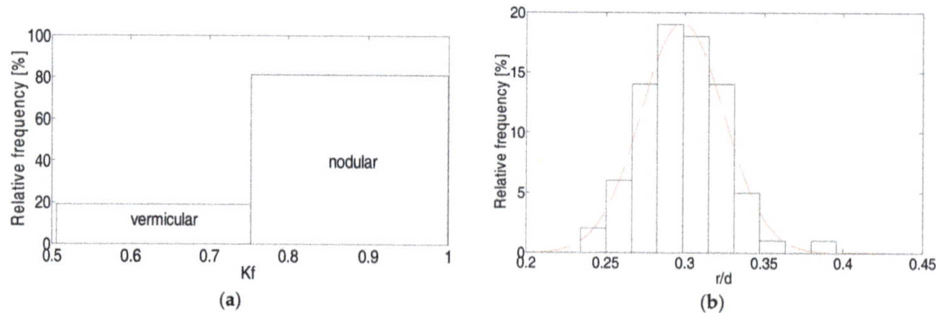

(a)   (b)

Figure 5. Geometric parameters for NCI: (**a**) histograms of K_f and (**b**) r/d ratio.

According to Figure 5a, it is possible to verify that the graphite nodules present a dominant circular shape, as most of the samples present K_f values between 0.75 and 1.0. However, some inclusions also present a vermicular shape. Due to this accentuated characteristic, in this work, the homogenization procedure using synthetic geometry will be developed considering the graphite as a circular shape. The r/d parameter was 0.3 ± 0.01, as indicated in Figure 5b, and the error was sufficiently low for the model representation of NCI. These parameters were coded into the numerical model to represent NCI.

4.2. Microstructure of NCI

The micrograms obtained from the laser confocal system were used to determine the NCI grain size. Figure 6 presents the microstructure of a sample of NCI after etching with a 3% nital solution. In Figure 6a, the ferritic-pearlitic structure of the sample is observed. This structure is surrounding the nodular graphite, forming a so-called "bullseye," and it contains certain alloying elements, such as magnesium and silicon, and small quantities of other elements that do not act during the nucleation process. The presence of pearlite lamellae is due to the high cooling rate of the iron from the eutectoid temperature. Figure 6b depicts the matrix structure with the "bullseyes" and the mixture of fine lamellar pearlite and a ferrite matrix structure. Finally, Figure 6c presents the pearlite as a thin lamellar structure, which is the result of a high cooling rate after casting. Graphite is surrounded by perlite, with the presence of some ferrite [30].

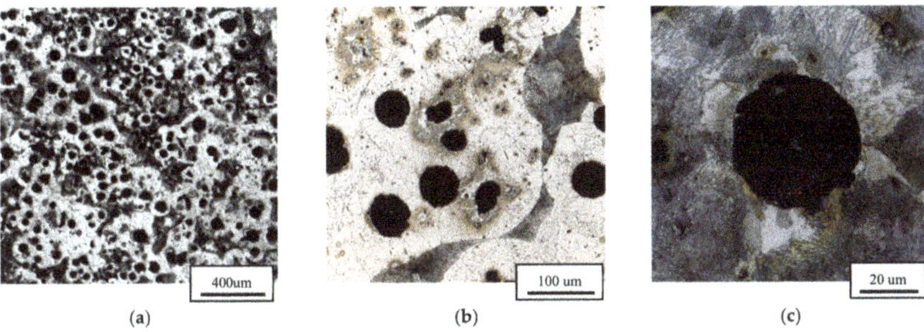

Figure 6. Metallographic structure of NCI: (**a**) ferritic-pearlitic structure, (**b**) "bullseye" structure, and (**c**) lamellar pearlite structure.

Table 2 summarizes the result for NCI that corresponds to the nodule diameter for a grain size number (G) of 4.62 (See ASTM E112-96). The average grain size of NCI obtained from the statistical count is illustrated in Figure 7.

Table 2. Experimental results from the laser micrographs.

Sample Comment	
ASTM grain size number G	4.62
Analysed area (mm^2)	6.55
Grains/unit area (N$^\circ$/mm^2) at 1×	13.31
Grains/unit area (N$^\circ$/mm^2) at 1×	5233.55
Average diameter (μm)	64.5
Elongation	0.99

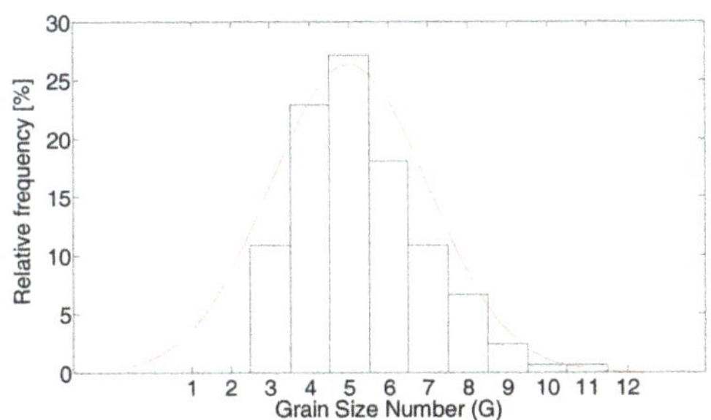

Figure 7. Grain size number (G) for GGG-40 obtained by the 3D laser microscopic system.

From Figure 7, the nodular graphite size indicates a value of G of around 5. In addition, ASTM E112-96 standard presents a relationship between G and the average diameter. For this study, the average diameter is approximately 64 ± 10 μm, which is in good agreement with related works [18,31,32].

4.3. Hardness Tests

The effective Young's modulus was determined from the Vickers micro-hardness tests, with the associated results presented via the histograms in Figure 8a,b for the Young's modulus of the matrix and the nodules, respectively.

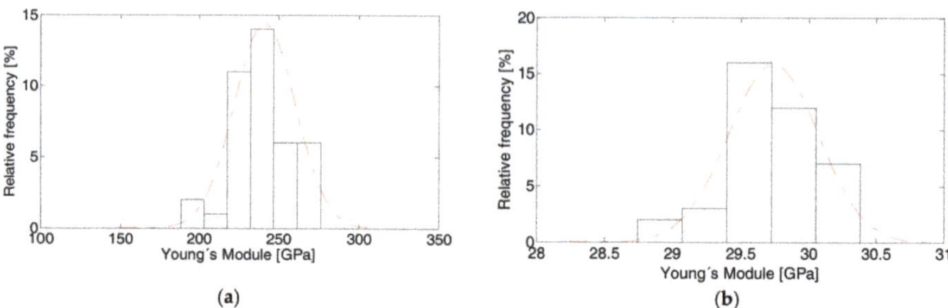

<table>
<tr><td>(a)</td><td>(b)</td></tr>
</table>

Figure 8. Histogram depicting the results for the Young's modulus of (**a**) the matrix and (**b**) the nodules.

The mean and standard deviations for the matrix and nodules are 241 ± 19 GPa and 29.17 ± 1.1 GPa, respectively. There is a significant difference between the Young's modulus of the matrix and that of the graphite nodules; the main reason for this is the low strength of the graphite. To ensure that the present results are valid, some studies regarding the determination of the Young's modulus for NCI are introduced and presented in Table 3.

4.4. Tensile Tests

Tensile tests were performed, and the stress versus strain curve is presented in Figure 9. The mechanical properties, including the yield stress ($\sigma_{0.2}$), ultimate tensile strength (σ_u), Young's modulus (E), and Poisson's ratio (ν), were obtained from the experimental curve, and they are reported in Table 3. The Young's modulus was 171 GPa, averaged over the four tests. The values of the modulus of elasticity that others obtained for NCI are 187 GPa [33], 172 GPa [15], and 179 GPa [34], demonstrating good agreement with the present values.

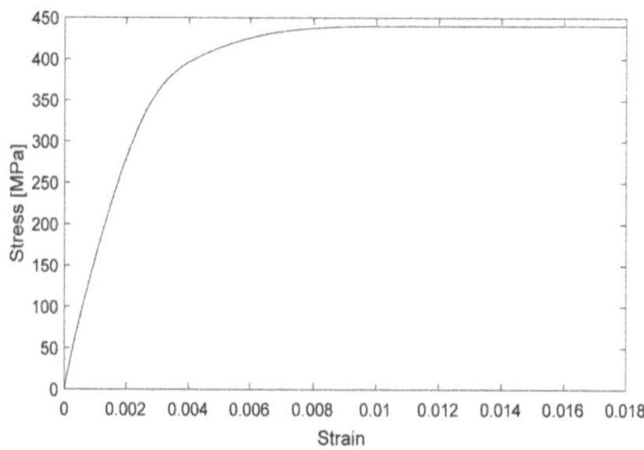

Figure 9. Stress-strain behavior of NCI determined from a tensile test.

Table 3. Experimental results from the tensile and micro-hardness testing.

Comment Results	Yield Stress $\sigma_{0.2}$ (MPa)	Ultimate Tensile Stress σ_{ut} (MPa)	Young's modulus E (GPa)	Comment Results	Ultimate tensile stress σ_{ut} (MPa)	Young's modulus E (GPa)
Experimental results	325	437	171	Matrix Nodules	532 –	241.2 29.2

4.5. Scanning Electron Microscope Equipped with EDX

During the formation of the nodular graphite in the casting process, additional elements, such as silicon and magnesium, are required to alter the solidification mechanism and chromium to improve the tensile strength. The chemical composition is presented in Table 4.

Table 4. The chemical composition of NCI (wt %).

C	Mg	Si	P	S	Cr	Mn	Ni	Cu	Fe
3.6	0.08	2.52	0.06	0.02	0.07	0.41	0.31	0.06	92.87

Nodular cast iron is formed by approximately 3.6% carbon, and it makes up 92.8% per mass of iron. The remaining chemical compounds were measured by SEM-EDX equipment (JEOL JSM 6610 SEM-EDX, Musashino, Tokyo, Japan). Despite obtaining information regarding all percentages of constituents (Table 4), it is important to note that this type of equipment is not able to detect small percentages of chemical constituents with good accuracy. For those constituents with high percentages (Fe, Si, and C), the values measured through the use of SEM-EDX equipment demonstrated good agreement with those provided by the foundry industry from which these samples were acquired. The high amounts of carbon are responsible for the formation of nodular graphite and for the lamellar perlite produced by the carbon remainder that did not form graphite during nucleation. According to Table 4, the high percentage of Si joined to the Mg explains the nodular graphite's round shape in the ferritic matrix. The small amounts of Cr and Cu account for the formations of the pearlite structure and the high tensile strength.

4.6. Computational Homogenization

4.6.1. Case I: Graphite Nodules Considered as Synthetic Geometry

Two-Dimensional Model

Figure 10 illustrates the convergence curve for E^*, where a wide dispersion for each computational test is evidenced. According to this graph, when a few nodules are considered, E^* varies considerably, indicating a positional dependence, which affects the effective properties significantly. It is possible to observe that as the number of nodules increases, E^* stabilizes for each set of samples. The elasticity modules, as well as the statistical data, are presented in Table 5.

In Table 5, E^* is the average value for each sample, $(E_{max} - E_{min})$ is the difference between the maximum and minimum values, (SD) is the standard deviation, (DVM) is the percentage deviation of the difference between maximum and minimum values with respect to the average, and (DM) is the percentage deviation of the standard deviation from the average value. Based on Table 5, it is possible to realize that E^* is reached with 35 nodules, and after this increment, the Young's modulus no longer varies significantly (less than 5%) except for particular points that affect the general result generated, as can be seen in Figure 11a. Figure 11a depicts the results for the E^* microstructures containing 35 nodules in each analysis; the maximum, minimum, average, and standard deviations of the group of samples tested are shown. Figure 11b presents the corresponding histograms of the results, where most of the analyses yielded values for E^* of 181 GPa. The value of E^* for the NCI obtained from the experimental data is 171 GPa.

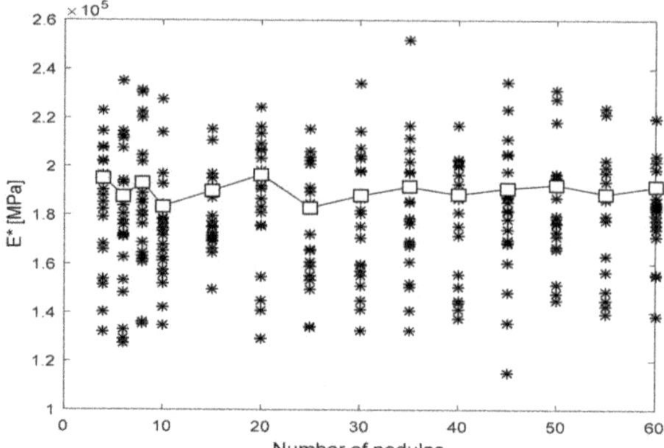

Figure 10. Convergence curve of the RVE. Values of the effective Young's modulus for samples with different random distributions of nodules.

Table 5. E^* for computational tests of the NCI microstructure with various random distributions of nodules.

N°	E^* (GPa)	$E^*_{max} - E^*_{min}$	SD	% (DVM) $\dfrac{E^*_{max} - E^*_{min}}{E^*_{media}} \times 100$	% (DM) $\dfrac{SD}{E^*_{media}} \times 100$
4	185.05	90.88	25.79	4.91	1.39
6	177.69	107.69	30.04	6.06	1.69
8	182.98	96.18	28.37	5.25	1.54
10	173.42	93.14	22.45	5.37	1.29
15	179.83	65.80	16.08	3.65	0.89
20	185.63	95.35	26.46	5.13	1.42
25	172.96	81.50	24.54	4.71	1.41
30	177.84	101.77	27.58	5.72	1.54
35	181.51	119.64	28.67	6.59	1.57
40	178.28	79.11	24.30	4.43	1.36
45	180.91	119.52	28.52	6.06	1.57
50	182.29	86.38	24.78	0.47	0.13
55	178.50	48.49	25.88	2.71	1.44
60	181.51	81.20	18.12	4.47	0.99

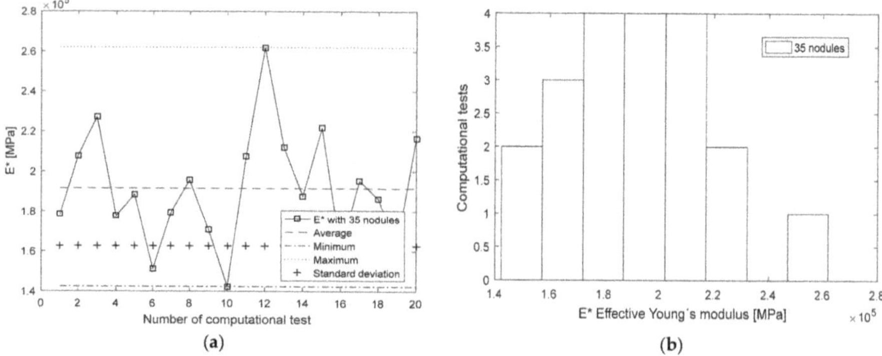

(a)   **(b)**

Figure 11. Statistical studies for average value: (**a**) E^* for 20 microstructures with 35 randomly distributed graphite nodules, and (**b**) the histogram of E^*.

The slight difference between the experimental and numerical results is approximately 8%. This difference is perhaps due to the 2D microstructural model employed and its simplification. According to Rodrígues et al. [14], a difference exists in the effective properties when using 2D and 3D RVEs. The authors claim that hydrostatic effects on the deformation of a 2D nodule are different from those effects that occur in a real 3D solid.

Three-Dimensional Model

In this work, the same strategy from [13] and [16] was used to study the effective property in an NCI RVE. Figure 12 presents the convergence curve of E^* for a 3D model. Furthermore, Table 6 lists the $(E_{max} - E_{min})$, the standard deviation, the percentage change in the difference between the maximum and minimum values with respect to sock (DVM) and the mean values of each of the samples for E^*, and the difference between the maximum and minimum values and the standard deviation (DM).

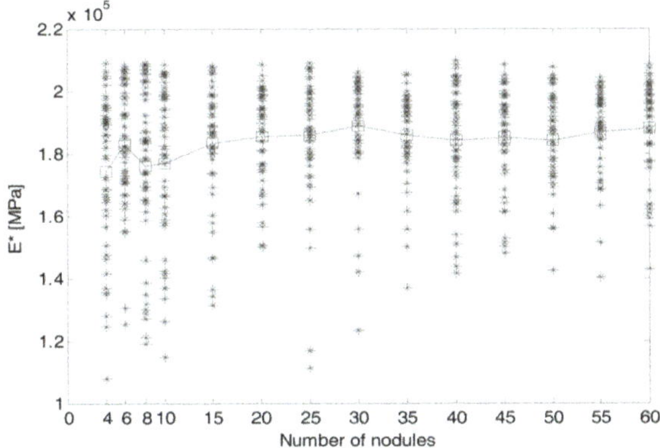

Figure 12. Convergence curve of the RVE. Values of the effective Young's modulus for samples with different random distributions of nodules.

Table 6. The E^* for computational test of the NCI microstructure with various random distribution of nodules.

N°	E (GPa)	$E^*_{max} - E^*_{min}$	SD	% (DVM) $\frac{E^*_{max} - E^*_{min}}{E^*_{media}} \times 100$	% (DM) $\frac{SD}{E^*_{media}} \times 100$
4	171.47	94.56	23.41	5.51	1.36
6	180.98	76.86	18.22	4.24	1.00
8	174.64	83.96	24.42	4.80	1.39
10	175.22	87.35	23.07	4.98	1.31
15	181.03	71.68	18.18	3.95	1.00
20	183.23	54.48	13.70	2.97	0.74
25	183.52	90.96	18.72	4.95	1.02
30	186.36	77.32	15.85	4.14	0.85
35	183.69	64.01	12.89	3.48	0.72
40	181.91	63.63	16.64	3.49	0.91
45	188.04	41.41	9.87	2.20	0.52
50	183.91	48.67	12.51	2.64	0.68
55	185.59	49.91	11.75	2.68	0.63
60	186.72	64.69	16.37	3.46	0.87

It can be observed that an E^* is reached for 30 inclusions, where there are increases of less than 5%. Figure 13a presents the result for E^* for an RVE with 30 random distributions of inclusions, while Figure 13b presents the corresponding histograms of the results, where it can be appreciated that the data for E^* were 186 GPa.

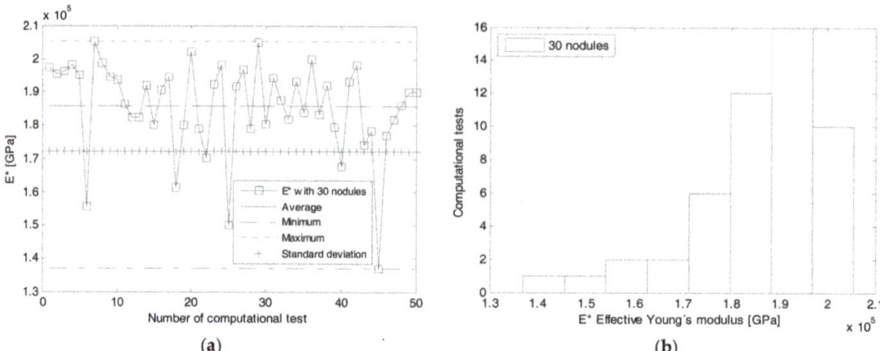

Figure 13. Statistical studies for average values: (**a**) E^* of material for 50 microstructures with 30 randomly distributed graphite nodules, and (**b**) the histogram of E^*.

Now, for the 3D analysis, it can be seen that the Young's modulus was 186 GPa due to the hydrostatic effects that generate a higher stiffness in the material.

4.6.2. Case II: Graphite Nodules Considered as Real Geometry

The homogenization was performed through the use of the real geometry of the NCI microstructure obtained by micro-CT acquisition. The results of the homogenization process are presented in Figure 14. It can be observed that as the number of inclusions increases, the effective Young's modulus is stabilized, reaching values of 175 GPa for an RVE with 30 inclusions. The RVE size in this case was 1.5 mm^2. The present methodology demonstrated good accuracy for determining the effective Young's modulus in relation to the values obtained with the experimental results, where the difference is approximately 8%. The error in the Young's modulus value can be justified by the 2D model approach. Despite considering the real geometry of the graphite, a difference of 8% remains between the values of the effective Young's modulus obtained experimentally and numerically. In this sense, perhaps a 3D model that takes into account the real geometry should be more realistic, thereby minimizing this difference. Table 7 summarizes the results for the effective Young's modulus obtained through homogenization (real and synthetic geometries) and the experimental procedure and those available in the literature.

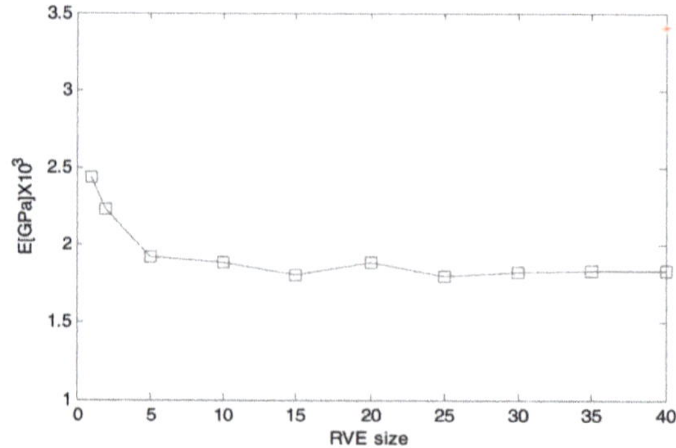

Figure 14. E^* for NCI.

In this work, three numerical models for determining the effective Young's modulus were considered. Table 7 summarizes the results for the effective Young's modulus obtained for each numerical model, the Young's modulus obtained by tensile test. Considering the approach of the graphite nodule as a synthetic geometry, both 2D and 3D models resulted in values greater than those obtained experimentally and those available in the literature. However, the 2D model that considers the real geometry of a graphite nodule resulted in good agreement with the values obtained experimentally and with [33]. In this sense, it may be possible to claim that those models that consider the real geometry of graphite nodules are more accurate for determining the effective properties.

Table 7. Results.

Model	E^* (GPa)
Synthetic 2D	181
Synthetic 3D	186
Real 2D	175
Experimental	171

5. Conclusions

In this paper, a homogenization process for determining the effective Young's modulus of NCI was performed. A numerical strategy for the homogenization process was written for a 2D and a 3D elastic model using all features provided for the BEM. The homogenization process was performed considering the nodular graphite modeled with synthetic and real geometries. Periodic boundary conditions were imposed to increase the convergence to reach an RVE. The real geometry of the microstructure was obtained from micro-CT images, and a special subroutine was developed to generate the BEM mesh. The graphite nodules were always considered as inclusions, and their geometries were modeled as subregions. In this sense, experimental studies were also performed to identify NCI's mechanical properties and morphological structure. From the results of the homogenization process, it is possible to verify that the modeling of the graphite as inclusions instead of voids significantly increases the accuracy. Regarding the graphite nodules, two different approaches were considered. The first one used the real geometry in which the BEM mesh was built based on micro-CT images of NCI. In the other approach, the graphite nodules were considered with a synthetic geometry; that is, the graphite shapes were considered to be perfectly round. The numerical results computed by the two approaches had nearly the same accuracy, whereas the modeling considering the real geometry of the graphite had higher computational efficiency than the model considering the synthetic geometry. The authors are, thus, devoting efforts to 3D BEM modeling considering the real geometry of NCI.

Author Contributions: A.B. and C.A. contributed with the numerical modelling and the experimental tests for the material characterization. A.M. and R.L. contributed with micro CT test.

Acknowledgments: The authors thank Capes and FAPDF for financial support.

Conflicts of Interest: The authors declare no conflict of interest.

References

1. Nguyen, V.P.; Stroeven, M.; Sluys, L.J. Multiscale Continuous and Discontinuous Modeling of Heterogeneous Materials: A Review on Recent Developments. *J. Multiscale Model.* **2011**, *3*, 1–42. [CrossRef]
2. Lee, K.; Ghosh, S. A microstructure based numerical method for constitutive modeling of composite and porous materials. *Mater. Sci. Eng. A* **1999**, *272*, 120–133. [CrossRef]
3. Héripré, E.; Dexet, M.; Crépin, J.; Gélébart, L.; Roos, A.; Bornert, M.; Caldemaison, D. Coupling between experimental measurements and polycrystal finite element calculations for micromechanical study of metallic materials. *Int. J. Plast.* **2007**, *23*, 1512–1539. [CrossRef]

4. Yue, X.; Weinan, E. The local microscale problem in the multiscale modeling of strongly heterogeneous media: Effects of boundary conditions and cell size. *J. Comput. Phys.* **2007**, *222*, 556–572. [CrossRef]
5. Patil, R.U.; Mishra, B.K.; Singh, I.V. A new multiscale XFEM for the elastic properties evaluation of heterogeneous materials. *Int. J. Mech. Sci.* **2017**, *122*, 277–287. [CrossRef]
6. Pundale, S.H.; Rogers, R.J.; Nadkarni, G.R. Finite Element Modeling of Elastic Modulus in Ductile Irons: Effect of Graphite Morphology. *J. Am. Foundrymen's Soc. Trans.* **1998**, *106*, 99–105.
7. Achenbach, J.D.; Zhu, H. Effect of interfacial zone on mechanical behavior and failure of fiber-reinforced composites. *J. Mech. Phys. Solids* **1989**, *37*, 381–393. [CrossRef]
8. Zhu, H.; Achenbach, J.D. Effect of Fiber-Matrix Interphase Defects on Microlevel Stress States at Neighboring Fibers. *J. Compos. Mater.* **1991**, *25*, 224–238. [CrossRef]
9. Ghosh, S.; Moorthy, S. Elastic-plastic analysis of arbitrary heterogeneous materials with the Voronoi Cell finite element method. *Comput. Methods Appl. Mech. Eng.* **1995**, *121*, 373–409. [CrossRef]
10. Yao, Z.; Kong, F.; Wang, H.; Wang, P. 2D simulation of composite materials using BEM. *Eng. Anal. Bound. Elem.* **2004**, *28*, 927–935. [CrossRef]
11. Huang, Q.; Zheng, X.; Yao, Z. Boundary element method for 2D solids with fluid-filled pores. *Eng. Anal. Bound. Elem.* **2011**, *35*, 191–199. [CrossRef]
12. Gitman, M.; Askes, H.; Sluys, L.J. Representative volume: Existence and size determination. *Eng. Fract. Mech.* **2007**, *74*, 2518–2534. [CrossRef]
13. Buroni, F.C.; Marczak, R.J. Effective properties of materials with random micro-cavities using special boundary elements. *J. Mater. Sci.* **2008**, *43*, 3510–3521. [CrossRef]
14. Rodríguez, F.J.; Dardati, P.M.; Godoy, L.a.; Celentano, D.J. Evaluación de propiedades elásticas de la fundición nodular empleando micromecánica computacional. *Rev. Int. Métodos Numéricos para Cálculo y Diseño en Ing.* **2014**, *31*, 1–15. [CrossRef]
15. Fernandino, D.O.; Cisilino, A.P.; Boeri, R.E. Determination of effective elastic properties of ferritic ductile cast iron by computational homogenization, micrographs and microindentation tests. *Mech. Mater.* **2015**, *83*, 110–121. [CrossRef]
16. Zohdi, T.I. Statistical ensemble error bounds for homogenized microheterogeneous solids. *ZAMP* **2005**, *56*, 497–515. [CrossRef]
17. Carazo, F.D.; Giusti, S.M.; Boccardo, A.D.; Godoy, L.A. Effective properties of nodular cast-iron: A multi-scale computational approach. *Comput. Mater. Sci.* **2014**, *82*, 378–390. [CrossRef]
18. Hütter, G.; Zybell, L.; Kuna, M. Micromechanisms of fracture in nodular cast iron: From experimental findings towards modeling strategies-A review. *Eng. Fract. Mech.* **2015**, *144*, 118–141. [CrossRef]
19. Brebbia, C.; Dominguez, J. *Boundary Elements An Introductory Course*, 2nd ed.; McGraw-Hill: New York, NY, USA, 1987.
20. Aliabadi, M.H. *The Boundary Element Method*, 1st ed.; Wiley: London, UK, 2002.
21. Nguyen, V.D.; Béchet, E.; Geuzaine, C.; Noels, L. Imposing periodic boundary condition on arbitrary meshes by polynomial interpolation. *Comput. Mater. Sci.* **2012**, *55*, 390–406. [CrossRef]
22. Larsson, F.; Runesson, K.; Saroukhani, S.; Vafadari, R. Computational homogenization based on a weak format of micro-periodicity for RVE-problems. *Comput. Methods Appl. Mech. Eng.* **2011**, *200*, 11–26. [CrossRef]
23. Burla, R.K.; Kumar, A.V.; Sankar, B.V. Implicit boundary method for determination of effective properties of composite microstructures. *Int. J. Solids Struct.* **2009**, *46*, 2514–2526. [CrossRef]
24. Nemat-Nasser, S.; Lori, M.; Datta, S.K. Micromechanics: Overall Properties of Heterogeneous Materials. *J. Appl. Mech.* **1996**, *63*, 561. [CrossRef]
25. Zohdi, T.I.; Wriggers, P.; Huet, C. A method of substructuring large-scale computational micromechanical problems. *Comput. Methods Appl. Mech. Eng.* **2001**, *190*, 5639–5656. [CrossRef]
26. Roula, A.; Kosnikov, G.A. Manganese distribution and effect on graphite shape in advanced cast irons. *Mater. Lett.* **2008**, *62*, 3796–3799. [CrossRef]
27. Ortiz, J.; Cisilino, A.; Otegui, J. Effect of Microcracking on the Micromechanics of Fatigue crack Growth in Austempered Ductile Iron. *Fatigue Fract. Eng. Mater. Struct.* **2001**, *24*, 591–605. [CrossRef]
28. Oliver, W.C.; Pharr, G.M. Measurement of hardness and elastic modulus by instrumented indentation: Advances in understanding and refinements to methodology. *J. Mater. Res.* **2004**, *19*, 3–20. [CrossRef]
29. Hollister, S.J.; Kikuehi, N. A comparison of homogenization and standard mechanics analyses for periodic porous composites. *Comput. Mech.* **1992**, *10*, 73–95. [CrossRef]

30. Bradley, W.L.; Srinivasan, M.N. Fracture and fracture toughness of cast irons. *Int. Mater. Rev.* **1990**, *35*, 129–161. [CrossRef]
31. Konečná, R.; Nicoletto, G.; Bubenko, L.; Fintová, S. A comparative study of the fatigue behavior of two heat-treated nodular cast irons. *Eng. Fract. Mech.* **2013**, *108*, 251–262. [CrossRef]
32. Steglich, D.; Brocks, W. Micromechanical modelling of damage and fracture of ductile materials. *Fatigue Fract. Eng. Mater. Struct.* **1998**, *21*, 1175–1188. [CrossRef]
33. Zhang, K.; Bai, J.; François, D. Ductile fracture of materials with high void volume fraction. *Int. J. Solids Struct.* **1999**, *36*, 3407–3425. [CrossRef]
34. Zybell, L.; Selent, M.; Hübner, P.; Kuna, M. Overload Effects During Fatigue Crack Growth in Nodular Cast Iron-simulation with an Extended Strip-yield Model. *Procedia Mater. Sci.* **2014**, *3*, 221–226. [CrossRef]

Article

Estimating the Effective Elastic Parameters of Nodular Cast Iron from Micro-Tomographic Imaging and Multiscale Finite Elements: Comparison between Numerical and Experimental Results

Andre Pereira [1,*], Marcio Costa [1,2], Carla Anflor [3], Juan Pardal [1] and Ricardo Leiderman [2,*]

[1] Engineering School, Fluminense Federal University, Rua Passo da Patria 156, Niterói 24210-240, Brazil; marciojairo.eng@gmail.com (M.C.); juanmanuelpardal@yahoo.com.br (J.P.)
[2] Institute of Computing, Fluminense Federal University, Rua Passo da Pátria 156, Niterói 24210-240, Brazil
[3] Group of Experimental and Computational Mechanics, University of Brasília, Gama 72.405-610, Brazil; anflorgoulart@gmail.com
* Correspondence: andremaues@id.uff.br (A.P.); leider@ic.uff.br (R.L.); Tel.: +55-21-98242-2070 (A.P.)

Received: 1 August 2018; Accepted: 4 September 2018; Published: 5 September 2018

Abstract: Herein, we describe in detail a methodology to estimate the effective elastic parameters of nodular cast iron, using micro-tomography in conjunction with multiscale finite elements. We discuss the adjustment of the image acquisition parameters, address the issue of the representative-volume choice, and present a brief discussion on image segmentation. In addition, the finite-element computational implementation developed to estimate the effective elastic parameters from segmented microstructural images is described, indicating the corresponding computational costs. We applied the proposed methodology to a nodular cast iron, and estimated the graphite elastic parameters through a comparison between the numerical and experimental results.

Keywords: nodular cast iron; effective Young's modulus; computational homogenization; multiscale numerical methods; micro-CT; finite elements

---

1. Introduction

Prior to the popularity of steel in construction, cast irons were the most widely used materials for such applications. Although cast iron has been mostly replaced by other materials, continuous improvements in metal-foundry processes have recently stimulated its production, resurrecting it as a major engineering material. Among the several types of cast irons available today, nodular (ductile, spheroidal) cast iron, which was developed in the mid-twentieth century, holds the most promise with respect to mechanical properties. The graphite phase within the nodular cast-iron matrix is in the form of small nodules, the size, shape, and arrangement of which control its effective mechanical properties, as illustrated in Fragassa et al. [1].

Scanning techniques such as microscopy are commonly used to capture the microstructures of cast irons. Although optical microscopy (as well as scanning electron microscopy) provides high-resolution imaging, its results are limited to the free surface of a specimen. Originally, studies on cast iron involved two-dimensional (2-D) images and models [2–4]. The quantification of graphite nodules by these techniques is possible but poor, because they only provide 2-D information, which limits the interpretation of results. Therefore, such models cannot provide a complete picture of the three-dimensional (3-D) cast-iron microstructures.

In recent years, X-ray micro-computed tomography (micro-CT) has proven to be well suited for the 3-D investigation of several materials, becoming an attractive non-destructive characterization technique in materials science [5]. More recently, micro-CT has also been applied to cast irons [6–8], showing that tomographic images provide affordable means to accurately describe, in 3-D, the morphological complexity of the graphite-nodule shapes and arrangements in nodular cast irons. Although these studies represent significant contributions, their quantitative analyses were restricted to geometric measurements such as the graphite volume fraction, nodule sphericity, minimum bounding-sphere diameter, graphite particle-size distributions, etc.

There is also a research community investigating the mechanical properties of nodular cast irons by means of numerical simulations and multiscale methods [2–4,9–11]. Therefore, the last two decades have evidenced a major shift, from treating cast irons only at the macroscale level, to complex multiscale models. Based on these well-documented numerical procedures and models [2–4,9–11], finite elements (FE) and homogenization techniques have been shown to be appropriate strategies for predicting the effective elastic properties of cast irons. However, most published studies have been restricted to the use of 2-D images [2–4], or have synthetically represented the graphite nodules as perfect discs or spheres [9–11], statistically distributed within the matrix. Such idealizations have led to poor numerical models, as indicated by Chuang et al. [7], where the authors investigated the complexity of the graphite 3-D structure, and its spatial distribution within the alloy.

To the best of our knowledge, there are no published studies that have taken advantage of the accurate description of the internal microstructure provided by micro-CT, to numerically evaluate the elastic properties of nodular cast irons. Therefore, the main aim of this work is to combine advanced numerical simulations with micro-CT imaging, to accurately compute the homogenized elastic parameters of nodular cast irons.

In addition, there are no literature reports that comprehensively describe the mechanical behavior of graphite nodules. To illustrate this, in Andriollo and Hattel [11], the authors summarized the graphite elastic parameters considered in several literature reports, highlighting that they range from 0 (void) to 375 GPa, as shown in Table 1 of the reference. In the present study, we apply the developed methodology to estimate the elastic parameters associated with graphite nodules. To that end, several numerical simulations were performed varying the graphite-nodule elastic parameters and keeping those of the ferritic-matrix fixed. The graphite elastic parameters were then estimated through a comparison between experimental and numerical results.

2. Materials and Methods

In this section, the nodular cast iron investigated in this study is introduced, and the methods used to obtain the homogenized elastic properties are discussed.

2.1. Nodular Cast Iron

The term "nodular cast iron" refers to a family of ternary alloys, formed by iron, carbon (4% maximum), and silicon (1.7 to 2.8%). This type of cast iron is characterized by the presence of carbon, in the form of spheroidal graphite, in a ferritic or pearlitic matrix. The graphite nodulation is achieved by adding a specific amount of magnesium to the molten metal. The resulting spheroidal graphite provides the nodular cast iron with better properties when compared to other cast irons, as discussed in Fragassa et al. [12]. The properties of nodular cast iron can be controlled through the cooling rate, addition of other alloying elements, casting method, and heat treatment, which may influence the graphite shape. The effective properties of these materials vary according to the graphite-nodule proportions and shapes. For these and other reasons, there is a need to characterize this material at the microstructural level.

The nodular cast iron used in this investigation was ASTM A536 (Metalrens, Minas Gerais, Brazil) grade 60-40-18 (the grade sequentially indicates the tensile strength (ksi), yield strength (ksi), and percent elongation at failure) ductile cast iron with a fully ferritic matrix, corresponding to the equivalent grade GGG40 for the DIN 1693 standard. The manufacturer was Metalrens, Ltd. [13]. Its mechanical behavior was characterized by tensile testing, and by digital and computational procedures consisting of micro-CT and multiscale homogenization with FE, respectively. The detailed test procedures are described in the following.

2.2. Experimental Tensile Testing

Tensile tests were performed according to the general guidelines given by the ASTM E8/E8M-09 Standard [14]. Four specimens were tested between parallel plates on an Instron 8801 test machine (Instron, Glenview, IL, USA) [100 kN (22,500 lbf)]. During testing, the specimen deflection increased at a constant rate, and was measured with a 2620-601 dynamic gauge extensometer (Instron, Glenview, IL, USA).

A single value was used for the elastic modulus obtained through the stress-strain curves. Since graphite sometimes exhibits nonlinear mechanical behavior, the use of a single-value elastic modulus could be questioned. Hence, the constitutive Young's modulus was identified as the slope of the experimental stress-strain curve at the beginning of the test, as indicated in Figure 1. The average initial tangent modulus obtained from several tests was $E = 181.9$ GPa.

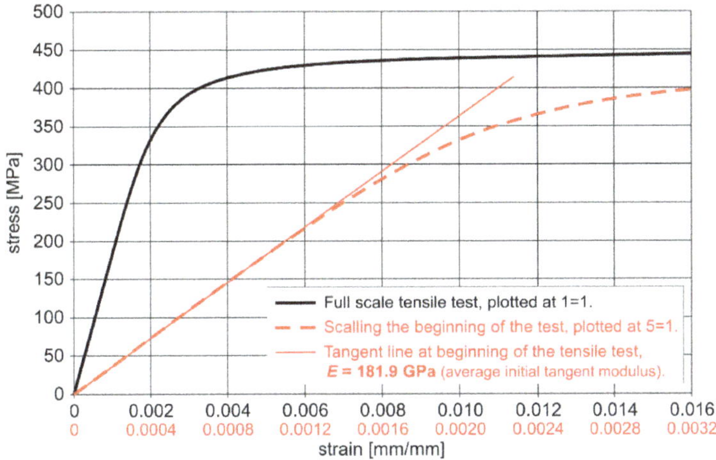

Figure 1. Stress-strain curve, with Young's modulus measurement, for a nodular cast iron sample.

2.3. X-Ray Micro-Tomography, Image Reconstruction, and Image Segmentation

X-ray micro-computed tomography (micro-CT), depicted in Figure 2, was conducted using a laboratory-based ZEISS VersaXRM-510 (ZEISS, Jena, Germany) high-resolution scanner system, with a minimum voxel size of 0.1 μm (resolution), and a maximum power output and voltage of 10 W and 160 kV, respectively. TXM Controller software (or the Scout-and-scan Control System, both by Zeiss, version 11.1.8043, Jena, Germany) was used to perform the scanning.

To obtain the tomographic images, the sample dimensions were specially chosen to ensure adequate X-ray transmission for high-resolution radiographs. Since cast irons are generally commercialized in the form of large blocks, small prismatic samples were extracted from a larger block, and machined to the $7 \times 8 \times 24$ mm^3 dimensions as illustrated in Figure 2a.

In our system, the sample to be imaged is positioned between an X-ray source and detector, as schematically represented in Figure 2b. The system generates X-rays with a cone-beam geometry. The detector registers the energy transmitted through each X-ray beam, which is proportional to the sample-density spatial distribution, thus generating a radiograph. Computed tomography (CT) is based on the repetitive radiographic sectioning of a sample, where radiograms are sequentially acquired while the sample (or the source-detector system) is incrementally rotated, as schematically shown in Figure 2b,c. From the collection of acquired radiograms, cross-sectional images of the sample are computed using tomographic-reconstruction algorithms, producing a stack of 2-D raw grey-scale horizontal-image slices. The 3-D digital object can then be visualized from the 2-D grey-scale image stack, as indicated in Figure 2d. In micro-CT, the dimensions of the reconstructed voxels (i.e., a volumetric pixel element) are generally in the micrometer range. TXM Reconstructor software was used to perform the image reconstruction.

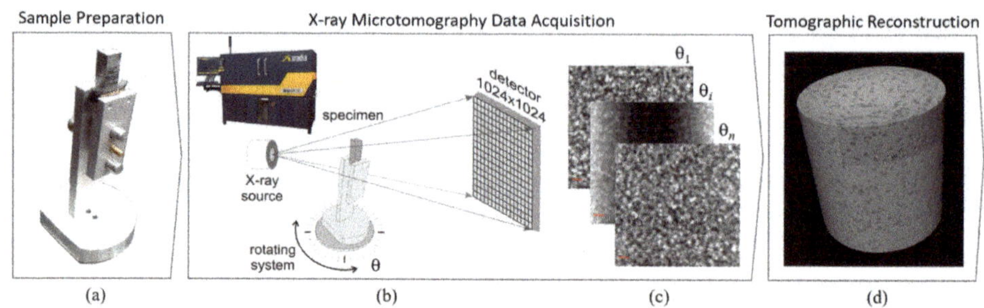

Figure 2. Schematic of the CT acquisition process: (**a**) sample preparation, dimensions of 7 mm × 8 mm × 24 mm; (**b**) setup of micro-CT scanner for image acquisition; (**c**) set of single radiographic projections; and (**d**) reconstructed volume with voxel size of 3 μm.

For high-quality micro-CT, all scans were conducted at an X-ray beam energy of 160 kV and a power of 10 W. The distance between the specimen and X-ray source, and between the specimen and detector, varied according to the target resolution. For the nodular cast iron sample under investigation, three micro-CTs were acquired, each with a different resolution. Optical magnifications of 0.4× and 4× were used for the lower (11 μm pixel size) and higher (both 3 μm and 1 μm pixel size) resolutions, respectively, in 1024 × 1024 pixel projection images. A different exposure time was set for each scan, to get intensity values ≥5000, for the best signal-to-noise ratio. Sets of 3200 projections (radiographs) were captured over a 360° sample rotation for each scan. Figure 3 shows the three regions of the sample that were imaged. The scanning parameters used to acquire the tomographic data, targeting three different resolutions, are shown in Table 1. The total scanning time for the 11 μm, 3 μm and 1 μm pixel size micro-CT was 2 h 40 min, 8 h 53 min and 26 h 40 min, respectively.

Table 1. Optimal scanning parameters for a 10.63 mm diagonal cast-iron block, at three different resolutions.

Resolution (Voxel Size)	Sample-Source Distance	Sample-Detector Distance	Optical Magnify-Cation	Filter (HE = High-Energy)	Beam Energy	Power	Exposure Time	Number of Projections
11 μm	30 mm	158 mm	0.4×	HE#4	160 kV	10 W	3.0 s	3200
3 μm	28 mm	35 mm	4.0×	HE#4	160 kV	10 W	10.0 s	3200
1 μm	26 mm	150 mm	4.0×	HE#6	160 kV	10 W	60.0 s	1600

The image resolution plays a central role in the success of the subsequent analyses. The resolution is typically selected based on two opposing criteria: (i) it should be high enough to resolve the microstructure, and (ii) it should be low enough for selection of a representative volume whose corresponding image size is treatable in the numerical simulations. In the current study, images of up to 400 × 400 × 400 voxels were considered as treatable. The issue of imaging-resolution selection is illustrated in the following example.

In Figure 3b–d, three micro-CT slices of the same nodular cast iron sample are shown. In Figure 3b, the sample was imaged with a resolution of 11 μm, while in Figure 3c,d it was imaged with a resolution of 1 μm and 3 μm, respectively. All the outer circles represented in the figures are of

1024 pixel diameter. Visual inspection showed that the 3 μm resolution was sufficient to resolve the microstructure, since the graphite-grain contours in Figure 3d could be easily distinguished, as well as in the three-dimensional view of Figure 3a. The 1 μm resolution (of course) was also sufficient to resolve the microstructure. In fact, the graphite-grain contour details in Figure 3c are superior to those in Figure 3d, although this did not impact the numerical simulations significantly. On the other hand, the 11 μm resolution (Figure 3b) was insufficient to resolve the microstructure. In addition, as indicated by Figures 4 and 5, for the 3 μm resolution, a 400 × 400 × 400 voxel cube would contain dozens, if not hundreds, of nodules, while, for the 1 μm resolution, a 400 × 400 × 400 voxel cube would contain only a few nodules. Further, for the 3 μm resolution, any 400 × 400 × 400 voxel cube would have approximately the same microstructure, being thus a representative volume, while, for the 1 μm resolution, 400 × 400 × 400 voxel cubes can potentially have very different microstructures. It can be concluded that the resolution of 3 μm is suitable for the present case, since it met both the criteria listed above.

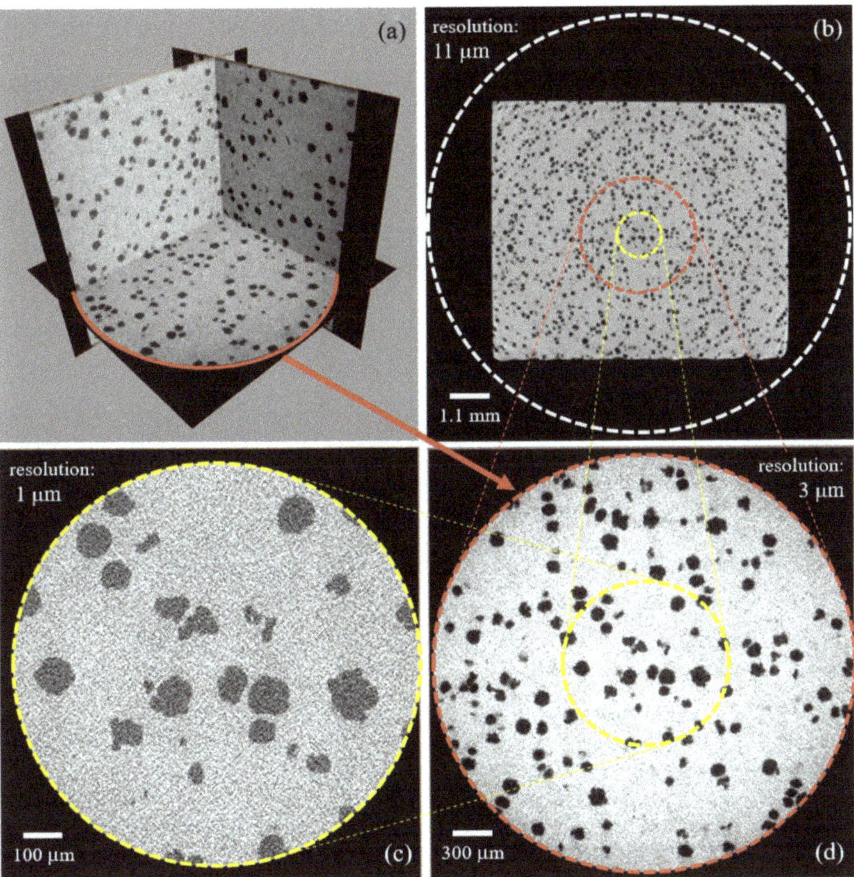

Figure 3. Three micro-CT (equivalent slices) obtained from the same nodular cast iron sample: (**a**) normal cross-sections at 3 μm resolution; (**b**) tomogram at 11 μm resolution; (**c**) 1 μm resolution; and (**d**) 3 μm resolution. The values for the corresponding acquisition parameters are in Table 1.

Direct volume rendering is a convenient way to explore micro-CT data. This method allows the analyst to study in detail the compositional variation in terms of connectivity, distribution, and relative density. 3-D image-analysis software (AVIZO, version 8.1, FEI, Massachusetts, MA, USA) was used for visualization, image processing, and segmentation. As described in Chuang et al. [7], the occurrence of coral-tree-like morphology that can span several-hundred microns in the iron matrix, was identified in the nodular cast-iron sample.

To perform complex measurements and to generate the model-mesh, it requires the elimination of noise and artefacts. Initially, the full volume represented by the stack of 1024 slices of 1024 × 1024 pixels, was cropped by discarding approximately the first-100 slices at the top and the last-100 slices at the bottom. This was necessary because the X-ray cone-beam geometry typically generates artefacts in those regions. Next, a denoising filter was applied to improve the reconstructed-image quality. There are several different filters available for this purpose, and the non-local means image-smoothing filter that removes high-frequency noise, was used in this analysis. Then, different phases of the object were identified and separated, through a procedure known as segmentation. In the context of digital material science, the main goal of image segmentation is to identify (and label) all the distinct phases in the material-sample digital image. A sophisticated and robust image-segmentation technique known as "watershed algorithm" was used. This technique involves placing seeds in different regions of the image and growing them until they reach a "barrier", determined by the image gradients. For image filtering and segmentation, a commercially available image-processing package (AVIZO) was used, which is capable of rapidly handling 1024 × 1024 × 1024 voxel images using an ordinary personal computer (PC). Illustrative results of filtering and segmentation for the nodular cast iron are depicted in Figure 4.

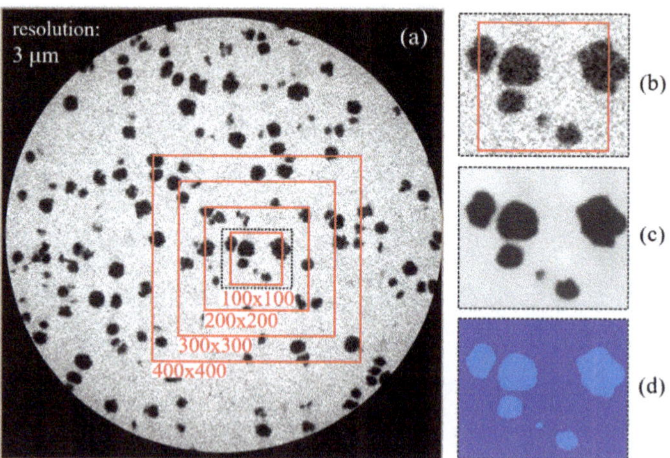

Figure 4. (**a**) A full slice of the micro-CT at 3 μm resolution, and four regions of interests with 100 × 100, 200 × 200, 300 × 300, and 400 × 400 pixels. Zoomed-in on a small region to show (**b**) the original image; (**c**) the image after filtering using a non-local mean filter; and (**d**) the watershed-algorithm segmented image.

Figure 4a shows a single slice of a non-processed micro-CT, zoomed-in on a small region (Figure 4b), and highlights the effect of applying a filter (non-local mean, Figure 4c) and performing a segmentation (watershed algorithm, Figure 4d). In Figure 4d, the ferritic matrix is in a dark-blue color and the graphite nodules are in light blue.

The next step was to create a set of regions of interest (ROI), as depicted in Figure 4a, for measurements and property estimates. For each ROI, the graphite/ferrite relative amount was calculated by dividing the total number of graphite-phase voxels by the total number of ROI voxels (since all voxels are of the same size). Generally, the larger the ROI, the more accurate the estimates are.

2.4. Homogenization with Finite Elements

The effective elastic parameters were computed using computational-homogenization techniques with the aid of the finite element method (FEM) [15]. The FEM is widely used in the context of computational solid mechanics, and is a numerical method mainly applied to solve (partial) differential equations. It is based on the weak (variational) formulation of a problem, and turns the original problem into a corresponding algebraic system through domain discretization. Computational homogenization [16], which is based on the premise of separation of scales, and on the concept of a representative volume element (RVE) [17], allows determination of the effective mechanical constitutive behavior of a composite material on a macroscopic scale, by modeling and investigating its microscopic features. This implies that with properly constructed models, a composite can be replaced by an effective homogeneous material for several practical means.

One of the main challenges in the present context is the generation of a FE mesh for a RVE of the actual material, to perform the numerical simulations. In the present study, the micro-CT 3-D images were converted into input data through a direct "voxel to finite element" transcription. Thus, the segmented micro-CT volumetric images were the starting point for the FE simulations. This involved the direct conversion of the 3-D digital-image voxel structure, where each voxel became an eight-node hexahedral element (with tri-linear shape functions). To use the FEM to analyze nodular cast irons and capture the graphite-nodule details, the element size needs to be significantly smaller than the nodule diameter, typically a few microns. With this type of discretization, millions of degrees of freedoms (DOFs) are required in the FEM model, to analyze a small material block. For example, a 3-D digital image composed of $400 \times 400 \times 400$ voxels (see Figure 4a) would require a system with approximately 193 million DOFs.

The large number of FEs in the model restricted the application of a standard FEM for solving the problem, due to limitations in computer-processing time and memory storage. Therefore, the micromechanical analyses performed in this study were carried out by means of an in-house written software, using C^{++} programming language and OpenMP extension for parallelism. To handle the large linear systems, a special technique known as element-by-element (EBE) implementation [18,19] was adopted. The implementation took advantage of the fact that all elements in the mesh were of the same size. In addition, as is common for the EBE implementation, the resulting linear system (that is not assembled) was solved with the aid of a preconditioned conjugate gradient (PCG) algorithm [20]. Furthermore, the computations for each element were performed in parallel. However, the tasks could not be distributed arbitrarily between the processors, since two processors working with two different elements sharing the same node, could attempt to update the same vector entry simultaneously, causing memory contention or dirty read/write entries. Hence, in the present implementation, the elements were divided into different "groups" (or "colors") such that elements of the same group did not share nodes. Within each group, the computations were then performed in parallel, although the groups were processed sequentially. For the 3-D regular mesh, where each voxel corresponded to an (cube) element, the elements were divided into 8 different groups.

Based on the mean-field homogenization schemes, there are three different numerical approaches for estimating the effective elastic parameters, as indicated in [21]. The first approach involves the application of a uniform strain-field component at the macroscale (a known average strain-field volume, typically unitary). This requires a microscale problem solution, with a prescribed equivalent displacement at the RVE boundary, and the posteriori computation of the average volume of the resulting stress-field using standard mathematical-averaging equations. Therefore, solving six different microscale problems (one for each uniform-strain component, corresponding to three uniaxial extensions and three simple-shear loadings), the averaging process leads to the full macroscopic stress field. The average volume of the resulting RVE stress field provides a specific column of the stiffness tensor (6×6 matrix representation). By relating the mean-stress and strain tensors, it is possible to obtain the homogenized constitutive relation at the macroscale. The second approach is similar and involves the application of a uniform stress-field component at the macroscale. (a known macroscopic average of stress-field volume). This requires a microscale problem solution, with a prescribed equivalent traction at the RVE boundary, and the posteriori computation of the mean strains. Several researchers have proved that imposing homogeneous strain-boundary conditions results in apparent properties close to the upper bound, while imposing homogeneous stress-boundary conditions generates estimates close to the lower bound. Hence, these two approaches provide two bounds for the effective elastic parameters. See Chevalier et al. [22], Harrison et al. [23], Niebur et al. [24], Rietbergen et al. [25], and Ulrich et al. [26] for examples of application with these two approaches. The third approach comes from the asymptotic multiscale homogenization, and consists of using the so-called periodic boundary conditions (PBC) [17]. The basic assumption behind the use of PBC is that the numerical model is a representative volume of an unbounded (statistically homogeneous) medium, i.e., a periodic microstructure is assumed throughout the medium. In the context of the FEM, PBC can be applied by assigning the same equation number to corresponding nodes at opposite sides, forcing the displacements at these corresponding nodes to be equal, and satisfying equilibrium. In this approach, the input is also a uniform strain-field component at the macroscale. See Nguyena et al. [21] for implementation details and Arns et al. [27], Garboczi and Day [28], Garboczi and Berryman [29], Makarynska et al. [30], and Roberts and Garboczi [31] for examples of application with this approach. The estimates provided by all three approaches described above get closer as the physical dimensions of the image (RVE) get larger in comparison to the microstructural characteristic length, and have been commonly used to estimate effective elastic parameters. However, the third approach naturally provides better approximations, and was therefore used in this study.

3. Results and Discussion

A realistic prediction of effective properties using a micromechanical model depends on whether the material behavior of each individual constituent and its geometry and topology can be accurately accounted for [32]. A realistic geometric representation of the models in the present study was achieved by using X-ray micro-CT. The heterogeneities were handled by direct numerical modeling of the composite structure, containing all the microstructural details, using the FEM. There were two well-distinguished constituents in the nodular cast-iron grade used in this work; a ferritic matrix and graphite nodules. No significant differences were found in the literature for the elastic properties of the ferritic matrix, and consensus values of 210 GPa for the Young's modulus and 0.3 for the Poisson's ratio were used in the computations. However, as mentioned previously, in Andriollo and Hattel [11], the authors showed that there is a wide range of graphite-nodule elastic parameters (Young's modulus and Poisson's ratio) found in the literature.

Therefore, to estimate the graphite-nodule Young's modulus (assuming the nodules to be homogeneous and isotropic) several computations were performed, covering a wide range of possible values (varying from 0 to 200 GPa), while keeping the Poisson's ratio fixed at 0.2225 (which is among the most commonly used values in the literature, i.e., 0, 0.2, 0.2225, or 0.3).

Since the representativeness of the numerical model plays an important role in the determination of the effective material properties (see, for example, Yu [32]), we considered the four ROIs shown in Figure 4a in our analysis, identifying the largest one (400^3 voxels) as a suitable RVE (as indicated in Figure 5).

Isotropic linear-elastic constitutive-material characteristics were assumed for both constituents of the nodular cast iron. However, although the constituents are, the composite material is not necessarily isotropic, due to natural anisotropies caused by heterogeneities. Therefore, the starting point to determine the effective elastic parameters using the present approach was to assume linear-elastic orthotropic material behavior. For this assumption, nine elastic parameters are required to define the constitutive behavior. Therefore, for each RVE, at least six elementary numerical tests were required at the microscale, by applying either uniform strains or uniform stresses at the macroscale, where the effective elastic properties were defined from spatial averages of fields over the RVE volume. If the three resulting longitudinal elastic moduli are the same, and the three resulting transversal moduli are also the same, this indicates that the material is indeed isotropic. This is a reasonable strategy for confirming isotropy, as well as ensuring that the considered RVE can be, in fact, assumed to be a homogeneous representative volume of the material. This is expected for a representative volume of nodular cast iron, since most metals are identified to be isotropic and homogeneous at the macroscale.

The four models (RVE candidates) shown in Figure 5, with sizes of 100^3, 200^3, 300^3, and 400^3 voxels, were simulated. The numerical models corresponded to the digital-sample central region, so that their centroids were identical (see Figure 5). To visually inspect the graphite-nodule distribution, the ferritic matrix was omitted from the figure, and only the rendered graphite regions are shown. Although the FE volumes are small, up to $(1.2 \text{ mm})^3$, in comparison to the laboratory tensile-test volume, they contained sufficient microstructural characteristics to be considered as potentially representative. Each analysis (one simulation for a unidirectional uniform-strain field with PBC) took approximately 2 min to run for a $100 \times 100 \times 100$ voxel input image (3 million DOFs), 20 min for a $200 \times 200 \times 200$ voxel input image (24 million DOFs), and 2 h for a $400 \times 400 \times 400$ voxel input image (193 million DOFs). In all the analysis, the PCG convergence criterion used was 10^{-8} for the relative norm of the residue vector. The estimates were performed using an ordinary desktop PC.

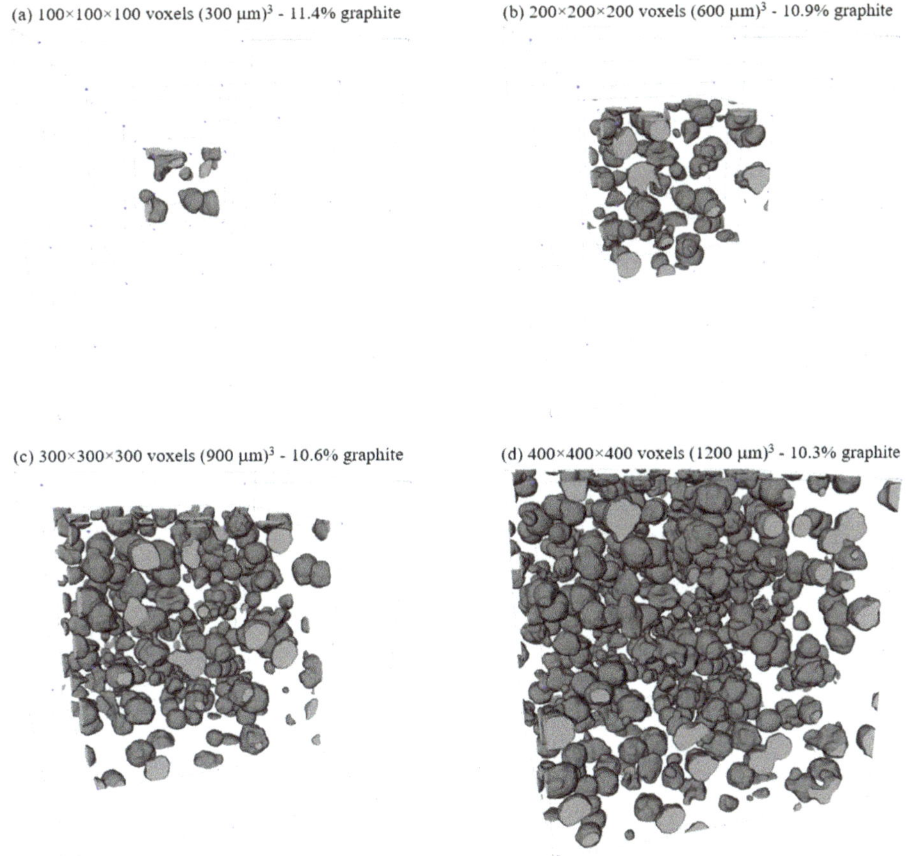

Figure 5. RVEs used in the homogenization process, generated from a micro-CT with a 3 μm voxel size, for the nodular cast iron sample (only the graphite nodules are rendered): RVE (**a**) with $100 \times 100 \times 100$; (**b**) with $200 \times 200 \times 200$ voxels; (**c**) with $300 \times 300 \times 300$ voxels; and (**d**) with $400 \times 400 \times 400$ voxels.

The results from the analysis of each FE model, by varying the graphite-nodule Young's modulus, are shown in Figure 6. From these plots, a convergence pattern in the RVE-size function can be observed. By using the largest FE model as the RVE of the nodular cast iron, a candidate value for the graphite Young's modulus was identified, by comparing the numerical result with the experimental result obtained from the slope of the tensile-test stress-strain curve shown in Figure 1 (181.9 GPa). In that sense, the predicted graphite Young's modulus value was found to be 39.7 GPa.

Next, assuming a graphite Young's modulus of 39.7 GPa, the respective effective elastic modulus was computed for each FE model. The results for the cast iron effective elastic-modulus convergence, as a function of the number of FEs used in each RVE, are plotted in Figure 7. More specifically, since the RVE was initially assumed to be orthotropic, the effective elastic modulus in the three orthogonal directions (E_x, E_y and E_z) are plotted. These results are also summarized in Table 2.

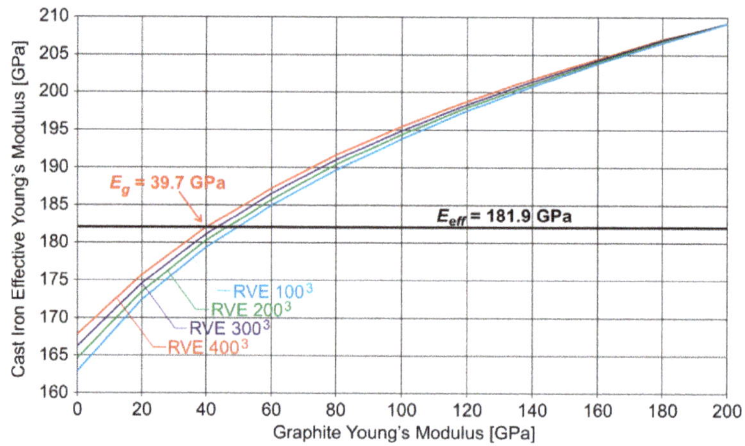

Figure 6. Results for cast iron effective elastic-modulus as a function of the graphite Young's modulus, for the RVE candidates shown in Figure 5.

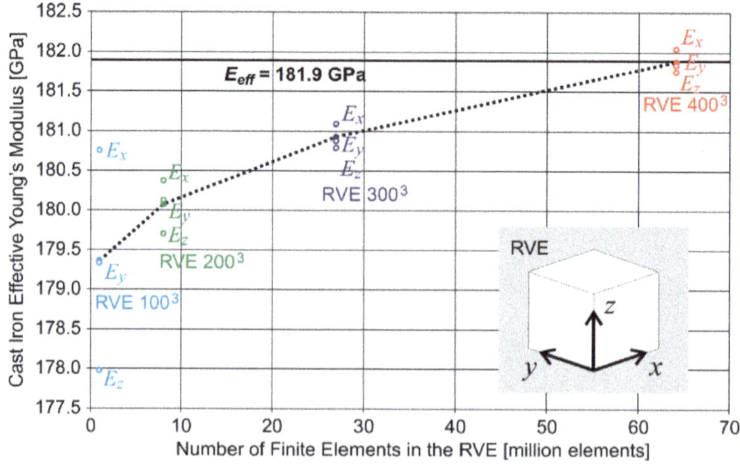

Figure 7. Results for cast iron effective elastic-modulus convergence as a function of the size of (or number of finite elements used in each) RVE, for the RVE candidates shown in Figure 5.

The analysis of Figure 7 (and Table 2) allows us to draw three main conclusions: (i) The largest RVE presents the largest effective elastic modulus. This is explained by the fact that the largest RVE contains the smallest amount of graphite (10.3%). Since the graphite Young's modulus is much smaller than the ferrite Young's modulus, it results that a smaller amount of graphite leads to a larger effective elastic modulus; (ii) The difference between E_x, E_y and E_z is smaller for the largest RVE. This is because the largest RVE contains a larger amount of nodules and, therefore, a larger variety of nodule shapes and a homogeneous nodule distribution, making it more representative of the nodular cast iron under investigation than the smaller RVEs. This is confirmed by the trend seen in the results that the smaller is the RVE the larger will be the difference between E_x, E_y and E_z; (iii) The nodular cast iron under investigation seems to be indeed isotropic, since the difference between E_x, E_y and E_z is very small. The results shown in Figure 7 deserve yet one last comment: It can be seen that E_y and E_z increase monotonically with the RVE size. The same not occurs with E_x that is larger for the 100^3 voxels volume than for the 200^3 voxels volume. This is an evidence that the volumes of 100^3 and 200^3 voxels shown in Figure 5 do not represent so accurately the cast iron.

Table 2. Cast iron effective elastic-modulus computed in each direction for each RVE candidate, assuming a graphite Young's modulus of 39.7 GPa.

Effective Elastic-Modulus	RVE 100 × 100 × 100	RVE 200 × 200 × 200	RVE 300 × 300 × 300	RVE 400 × 400 × 400
E_x (GPa)	180.757	180.376	181.090	182.043
E_y (GPa)	179.338	180.129	180.876	181.838
E_z (GPa)	177.976	179.700	180.790	181.752
Average (GPa)	179.357	180.068	180.919	181.878

4. Conclusions

In the present work, we described in detail a methodology to estimate the effective elastic parameters of nodular cast iron, using micro-computed tomography in conjunction with multiscale finite elements. We discussed the adjustment of the image acquisition parameters and addressed the issue of representative volume choice. We also presented a brief discussion on image segmentation and described the finite element computational implementation we developed to estimate the effective elastic parameters from segmented microstructural images. Our finite element implementation easily handled images of up to $400 \times 400 \times 400$ voxels (~2×10^8 DOFs). We then applied the proposed methodology to a nodular cast iron and specifically estimated the graphite elastic parameters through a comparison between numerical and experimental results, by trial and error. The graphite Young's modulus was estimated to be 39.7 GPa, which is within the range of values found in the literature. The numerical results also indicated that the considered volumes of 100^3 and 200^3 voxels (2.7×10^{-2} mm^3 and 2.16×10^{-1} mm^3, respectively) do not represent so accurately the cast iron, and that the cast iron is indeed isotropic at the macroscale.

Author Contributions: A.P. wrote the manuscript, performed the micro-CT and implemented the software to perform the computational homogenization; M.C. processed the micro-CT and implemented the software to perform the computational homogenization; C.A. performed the mechanical test and analyzed the data; J.P. analyzed the data; R.L. designed the software to perform the computational homogenization, analyzed the data and wrote the manuscript.

Funding: This research received no external funding.

Acknowledgments: This research was partially supported by CNPq (National Council for Scientific and Technological Development) Brazilian agency, grant number 404593/2016-0.

Conflicts of Interest: The authors declare no conflict of interest.

References

1. Fragassa, C.; Minak, G.; Pavlovic, A. Tribological aspects of cast iron investigated via fracture toughness. *Tribol. Ind.* **2016**, *38*, 1–10.
2. Carazo, F.D.; Giusti, S.M.; Boccardo, A.D.; Godoy, L.A. Effective properties of nodular cast-iron A multi-scale computational approach. *Comput. Mater. Sci.* **2014**, *82*, 378–390. [CrossRef]
3. Fernandino, D.O.; Cisilino, A.P.; Boeri, R.E. Determination of effective elastic properties of ferritic ductile cast iron by computational homogenization, micrographs and microindentation tests. *Mech. Mater.* **2015**, *83*, 110–121. [CrossRef]
4. Kasvayee, K.A.; Salomonsson, K.; Ghassemali, E.; Jarfors, A.E.W. Microstructural strain distribution in ductile iron; comparison between finite element simulation and digital image correlation measurements. *Mater. Sci. Eng. A* **2016**, *655*, 27–35. [CrossRef]
5. Landis, E.N.; Keane, D.T. X-ray microtomography. *Mater. Charact.* **2010**, *61*, 1305–1316. [CrossRef]
6. Mrzygłód, B.; Matusiewicz, P.; Tchórz, A.; Olejarczyk-Wożeńska, I. Quantitative Analysis of Ductile Iron Microstructure—A Comparison of Selected Methods for Assessment. *Arch. Foundry Eng.* **2013**, *13*, 59–63. [CrossRef]
7. Chuang, C.; Singh, D.; Kenesei, P.; Almer, J.; Hryna, J.; Huff, R. 3D quantitative analysis of graphite morphology in high strength cast iron by high-energy X-ray tomography. *Scr. Mater.* **2015**, *106*, 5–8. [CrossRef]
8. Yin, Y.; Tu, Z.; Zhou, J.; Zhang, D.; Wang, M.; Guo, Z.; Liu, C.; Chen, X. 3D Quantitative Analysis of Graphite Morphology in Ductile Cast Iron by X-ray Microtomography. *Metall. Mater. Trans. A* **2017**, *48*, 3794–3803. [CrossRef]
9. Zybell, L.; Hütter, G.; Linse, T.; Mühlich, U.; Kuna, M. Size effects in ductile failure of porous materials containing two populations of voids. *Eur. J. Mech. A Solids* **2014**, *45*, 8–19. [CrossRef]
10. Hutter, G.; Zybell, L.; Kuna, M. Micromechanisms of fracture in nodular cast iron: From experimental findings towards modeling strategies—A review. *Eng. Fract. Mech.* **2015**, *144*, 118–141. [CrossRef]
11. Andriollo, T.; Hattel, J. On the isotropic elastic constants of graphite nodules in ductile cast iron: Analytical and numerical micromechanical investigations. *Mech. Mater.* **2016**, *96*, 138–150. [CrossRef]
12. Fragassa, C.; Radovic, N.; Pavlovic, A.; Minak, G. Comparison of mechanical properties in compacted and spheroidal graphite irons. *Tribol. Ind.* **2016**, *38*, 49–59.
13. METALRENS 2017. Available online: http://www.metalrens.com.br/ (accessed on 8 December 2017).
14. *ASTM E8/E8M-09, Standard Test Methods for Tension Testing of Metallic Materials*; ASTM International: West Conshohocken, PA, USA, 2009. [CrossRef]
15. Guedes, J.M.; Kikuchi, K. Preprocessing and postprocessing for materials based on the homogenization method with adaptive finite element methods. *Comput. Meth. Appl. Mech. Eng.* **1990**, *83*, 143–198. [CrossRef]
16. Geers, M.G.D.; Kouznetsova, V.G.; Brekelmans, W.A.M. Multi-scale computational homogenization: Trends and challenges. *J. Comput. Appl. Math.* **2010**, *234*, 2175–2182. [CrossRef]
17. Pindera, M.J.; Khatam, H.; Drago, A.S.; Bansal, Y. Micromechanics of spatially uniform heterogeneous media: A critical review and emerging approaches. *Compos. Part B* **2009**, *40*, 349–378. [CrossRef]
18. Hughes, T.J.R.; Levit, I.; Winget, J. An element-by-element solution algorithm for problems of structural and solid mechanics. *Comput. Meth. Appl. Mech. Eng.* **1983**, *36*, 241–254. [CrossRef]
19. Erhel, J.; Traynard, A.; Vidrascu, M. An element-by-element preconditioned conjugate gradient method implemented on a vector computer. *Parallel Comput.* **1991**, *17*, 1051–1065. [CrossRef]
20. Shewchuk, J.R. *An Introduction to the Conjugate Gradient Method without the Agonizing Pain*; Technical Report for Carnegie Mellon University: Pittsburgh, PA, USA, 1994.
21. Nguyen, V.D.; Béchet, E.; Geuzaine, C.; Noels, L. Imposing periodic boundary condition on arbitrary meshes by polynomial interpolation. *Comput. Mater. Sci.* **2012**, *55*, 390–406. [CrossRef]
22. Chevalier, Y.; Pahr, D.; Allmer, H.; Charlebois, M.; Zysset, P. Validation of a voxel-based FE method for prediction of the uniaxial apparent modulus of human trabecular bone using macroscopic mechanical tests and nanoindentation. *J. Biomech.* **2007**, *40*, 3333–3340. [CrossRef] [PubMed]
23. Harrison, N.M.; McDonnell, P.F.; O'Mahoney, D.C.; Kennedy, O.D.; O'Brien, F.J.; McHugh, P.E. Heterogeneous linear elastic trabecular bone modelling using micro-CT attenuation data and experimentally measured heterogeneous tissue properties. *J. Biomech.* **2008**, *41*, 2589–2596. [CrossRef] [PubMed]

24. Niebur, G.L.; Feldstein, M.J.; Yuen, J.C.; Chen, T.J.; Keaveny, T.M. High-resolution finite element models with tissue strength asymmetry accurately predict failure of trabecular bone. *J. Biomech.* **2000**, *33*, 1575–1583. [CrossRef]

25. Rietbergen, B.; Weinans, H.; Huiskes, R.; Odgaardt, A. A new method to determine trabecular bone elastic properties and loading using micromechanical finite-element models. *J. Biomech.* **1995**, *28*, 69–81. [CrossRef]

26. Ulrich, D.; Rietbergen, B.; Weinans, H.; Ruegsegger, P. Finite element analysis of trabecular bone structure: A comparison of image-based meshing techniques. *J. Biomech.* **1998**, *31*, 1187–1192. [CrossRef]

27. Arns, C.H.; Knackstedt, M.A.; Pinczewski, W.V.; Garboczi, E.J. Computation of linear elastic properties from microtomographic images: Methodology and agreement between theory and experiment. *Geophysics* **2002**, *67*, 1396–1405. [CrossRef]

28. Garboczi, E.J.; Day, A.R. An algorithm for computing the effective linear elastic properties of heterogeneous materials: Three-dimensional results for composites with equal phase poisson ratios. *J. Mech. Phys. Solids* **1995**, *43*, 1349–1362. [CrossRef]

29. Garboczi, E.J.; Berryman, J.G. Elastic moduli of a material containing composite inclusions: Effective medium theory and finite element computations. *Mech. Mater.* **2001**, *33*, 455–470. [CrossRef]

30. Makarynska, D.; Gurevich, B.; Ciz, R.; Arns, C.H.; Knackstedt, M.A. Finite element modelling of the effective elastic properties of partially saturated rocks. *Comput. Geosci.* **2008**, *34*, 647–657. [CrossRef]

31. Roberts, A.P.; Garboczi, E.J. Computation of the linear elastic properties of random porous materials with a wide variety of microstructure. *Proc. R. Soc. Lond. A* **2002**, *458*, 1033–1054. [CrossRef]

32. Yu, W. A unified theory for constitutive modeling of composites. *J. Mech. Mater. Struct.* **2016**, *11*, 379–411. [CrossRef]

Article

Influence of Heat Treatment in the Microstructure of a Joint of Nodular Graphite Cast Iron when Using the Tungsten Inert Gas Welding Process with Perlitic Grey Cast Iron Rods as Filler Material

Francisco-Javier Cárcel-Carrasco *, Manuel Pascual-Guillamón, Fidel Salas-Vicente and Vicente Donderis-Quiles

Inst. Tecnología Materiales, Universitat Politècnica de València, 46022 Valencia, Spain; mpascual@mcm.upv.es (M.P.-G.); fisavi@doctor.upv.es (F.S.-V.); vdonderis@die.upv.es (V.D.-Q.)
* Correspondence: fracarc1@csa.upv.es (F.-J.C.-C); Tel.: +34-963-87-7000; Fax:+34-963-87-9459

Received: 9 December 2018; Accepted: 30 December 2018; Published: 7 January 2019

Abstract: The present article analyses the influence of preheating and a postweld heat treatment in the microstructure, mechanical properties and wear behaviour of a joint of nodular graphite cast iron when using the tungsten inert gas (TIG) welding process with perlitic grey cast iron rods as filler material. Data obtained from the tests and the microstructural study of the samples show that the absence of a postweld heat treatment and of preheating leads to the apparition of hard structures and a notable reduction in elongation. Preheating or annealing the weld avoid the presence of these hard structures and increase the ductile behaviour of the joint although at the cost of a further loss of mechanical strength. Wear rate was found to be higher at the weld bead than at the base metal, even when the hardness of both areas is the same.

Keywords: weldability; pre-heating; spheroidal graphite cast iron

1. Introduction

Cast irons are alloys of iron, carbon and silicon that favours the formation of graphite, whose carbon content, due to the fact that a high percentage of carbon can induce brittleness, is maintained below 4%. Their classification depends on its metallographic structure, according to the percentages of carbon and other alloying elements. Although these materials do not have the mechanical properties of steels, they are used in numerous applications such as: hydraulic valves, transmissions, gears, nuts, shafts, hydraulic components, pistons, guides or engine liners [1].

Grey cast irons have low mechanical characteristics due to the presence of graphite flakes, which act as discontinuities in the matrix, giving rise to the presence of stress concentrators. Because of the good castabilityand low price of grey cast irons, they are used extensively for ornamental objects, manhole covers, heat exchangers or bandstands

In contrast, in nodular cast irons the addition of Magnesium, Magnesium-ferrosilicon or Magnesium-Nickel favours the formation of graphite nodules instead of graphite flakes, improving the mechanical strength, toughness and ductility of the cast iron without losing the ease of the moulding ofgrey cast irons [2]. This type of cast iron is employed in pipes and fittings, automotive applications, agricultural machinery and general industrial equipment.

Malleable cast irons castings are another type of cast irons, although with less applications than grey or nodular cast irons. They are basically iron-carbon alloys with a high carbon content that has a graphite-free structure (white cast irons) in their as-cast condition but are afterwards subjected to a thermal treatment that leads to the dissolution of cementite and the apparition of graphite.

There are two types of malleable castings: White heart malleable cast iron (European), which are decarburized cast irons with graphite in the form of irregular nodules and black heart cast iron (American) whose final structure after the heat treatment of the white cast iron is formed mainly of pearlite or perlite and irregular nodules of graphite.

Welding cast iron pieces is not a common practice and usually it is limited to repair operations and not joining due to the apparition of martensite or fragile carbides during the cooling phase of the joint. Nevertheless and although joining of cast iron pieces remains a difficult issue, it is possible if some precautions are taken.

The weld is usually performed using Ni, Ni-Cu or Ni-Fe electrodes for the best performance, although if a lower cost is mandatory, low-carbon manganese steels for non-machinable joints or cast iron rods can be used.

The mechanical properties of a weld joint depend on the welding process and the filler material but also on the preheating temperature and on the duration and temperature of the postweld heat treatment [3]. These two heat treatments are optional but necessary for high carbon content alloys such as cast irons in order to reduce the cooling rate and avoid or reduce the apparition of hard structures at the heat affected zone (HAZ).

The aim of this study has been to evaluate the influence of preheating and subjecting the weld to a postweld heat treatment on the mechanical characteristics of a TIG weld of two nodular iron plates when a perlitic grey cast iron rod is used as filler material. The use of grey cast iron as filler material is justified by its lower cost when compared to Ni electrodes and the possibility of obtaining a machinable joint. Furthermore, the presence of hard microstructures is almost unavoidable in cast iron joints, even after an annealing when high cost Ni electrodes have been used [3–5] but it could be less difficult to get rid of them with a thermal treatment if cast iron rods are used.

The wear behaviour of the joint was also evaluated due to the fact that many tools fabricated using nodular cast iron are affected by wear (ploughs, gears, drums, automotive components, etc.) [6–8] and the changes in hardness and microstructure caused by the welding process will affect its wear resistance mainly due to the presence of hard structures of martensite and carbides.

2. Materials and Methods

2.1. Materials

A nodular cast iron was used as base material. Its chemical composition and mechanical characteristics are shown in Table 1 and the mechanical properties of this cast iron can be seen in Table 2.

Table 1. Composition of the base and the filler materials.

Nodular Cast Iron (Base Material)									
C	Mn	S	Ni	Cu	Si	P	Cr	Mo	Mg
3.61	0.045	0.007	0.02	0.026	2.8	0.02	0.03	<0.01	<0.032
Perlitic Grey Cast Iron (Filler Material)									
C	Mn	S	Ni	Cu	Si	P	Cr	Mo	Mg
2.5	0.4	0.01	-	-	1.1	0.09	-	-	-

As filler meal, an ER perlitic grey cast iron rod with a diameter of 4 mm whose composition and properties can be seen in Tables 1 and 2 was selected. This material has lower properties than the base material due to the effect of the graphite flakes and does not seem to be the best option to weld nodular cast irons. Nevertheless, it is important to know the microstructure and properties of the joint and how the best properties can be obtained in case no other suitable material is available.

Table 2. Mechanical properties of the base and the filler materials.

Mechanical Properties		Perlitic Grey Cast Iron	Nodular Cast Iron
Ultimate strength	MPa	325	420
Yield strength	MPa	305	340
Elongation	%	9	14
Young modulus	MPa	130,000	160,000
Hardness	HV	248	196

The microstructure of both materials can be seen in Figure 1. The nodular cast iron has a ferritic-perlitic matrix and the grey cast iron has a perlitic matrix with flake graphite type V (star-like).

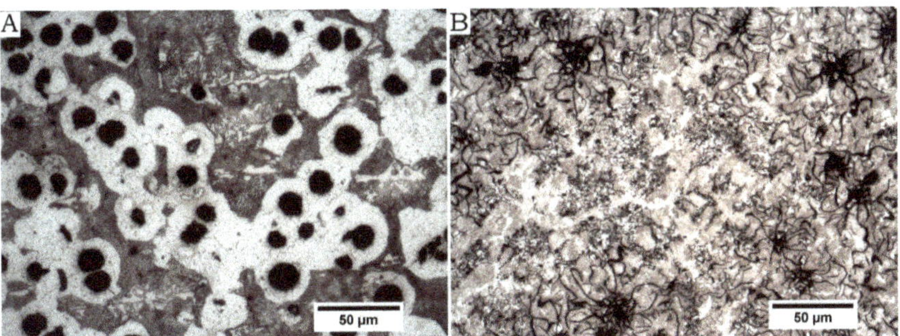

Figure 1. Microstructure of cast irons (**A**) Nodular cast iron (**B**) Perlitic grey cast iron.

2.2. Welding Processes

The plates were obtained by sand mould casting six plates of dimensions 170 mm ×50 mm × 6 mm. These plates where subsequently milled to dimensions 170 mm × 50 mm × 5 mm. The edges were prepared for welding with a single 30° bevel by means of a refrigerated cut with an adjustable band saw.

The joint was carried out using the TIG process, one of the techniques currently most used for elements whose thickness is less than 8 mm and for the root pass in parts of greater thicknesses in order to avoid a lack of penetration. The use of this technique is advisable in the welding of castings of spheroidal graphite of certain thicknesses that require previous preparation [9]. The welding parameters were: DC current between 120 and 130 A at 14 V, direct polarity and an Argon gas flow rate of 12 L/min. The net heat input was 455 J/mm.

The welds were done in two passes using a Fronius Trans TIG 1700 machine (a first root pass and a second filling pass, Fronius international, Wels, Austria), both with a circular movement of advance in the horizontal plane and from right to left. The angle of inclination of the tungsten electrode was between 70° and 80° in the advance direction while the angle of the filler rod was about 20° on the horizontal plane. A separation of 2 mm was maintained between both plates and the bevels were cleaned and deoxidized. This helps avoid undercuts and inclusions and ensure a good penetration.

Due to the difficulty in the process of welding cast irons and in order to avoid cracking due to the stresses generated in the cooling, the weld passes were made in lengths of approximately 30 mm that were peened while hot with a small peen hammer to relieve residual stresses. Another drawback of cast iron welding is the lack of fluidity of the molten metal, what causes porosity and complicates obtaining a good penetration.

As cast irons have a great tendency to produce fragile structures at the HAZ during the cooling of the joint, once finished, the weld was covered with a ceramic blanket to avoid rapid cooling [10].

The welded samples were divided into three groups depending on the exact welding process:

- Group 1: Without preheating or postweld heat treatment.

- Group 2: The coupon was annealed at 900 °C for 1 h and slowly cooled down inside the furnace.
- Group 3: The plates where preheated up to 450 °C before welding. The temperature was maintained around this value between passes.

2.3. Tests

Different test samples were obtained from each coupon:

- Five test pieces of 20 mm width and 100 mm length for tensile tests
- One sample for metallographic examination of the weld and microhardness measurement
- One sample for pin-on-disc tests.

The tensile tests were performed in an Instron universal testing machine (Instron, Norwood, MA, USA) model 4204 at a speed of 5 mm/min according to the UNE-EN 10002-1 standard for tensile tests at room temperature. SEM images were taken for each one of the surfaces of the breakthroughs at 3 kV in order to obtain information about the fracture.

For the micrographic examination, the specimens were grinded up to number 1000 sandpaper and subsequently polished using 3 and 1 μm diamond paste. Finally they were etched with Nital 3.

Microhardness was measured according to the UNE-EN 876 standard using an Innovatest 400Amicrohardness tester (Innovatest Europe BV, Maastricht, The Netherlands) with a load of 300 g and a dwell time of 10 s at the HAZ, the fusion line and the weld bead. The hardness corresponds to the mean value of 5 indentations.

To compare the wear behaviour of the weld bed with the wear behaviour of the HAZ and the base metal a pin-on-disk test was carried out for each one of the groups using a normal load of 10 N and a radius of 5 mm. A F-5210 steel ball of 5 mm diameter was used in the tests, that covered 50 m of distance at a rotational speed of 60 rpm. The circular path of the ball covered both the weld bead and the base metal. The surface was grinded up to 500 grit sandpaper before the tests.

After the pin-on-disc tests the response of the material was evaluated measuring the profile of the wear track using a Marh M2 profilometer (Microtest, Madrid, Spain) [11,12].

3. Results

Table 3 show the mean value of the results obtained from the tensile tests and Figure 2 shows the microhardness profile evolution from the base metal to the weld bead. Each hardness value corresponds to the mean value of 5 measurements.

Table 3. Mechanical characteristics of spheroidal graphite cast iron welding (A is the Elongation).

Group	Ultimate Strength (MPa)	Yield Strength (MPa)	A%
Without heat treatment	370 ± 18	330 ± 18	9
900 °C postweld annealing	320 ± 11	295 ± 11	12
450 °C preheating	335 ± 10	310 ± 10	10.5

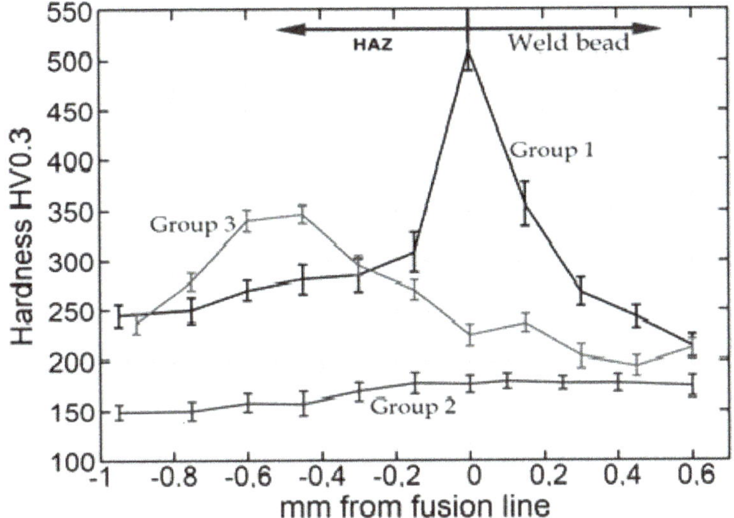

Figure 2. MicroVickers hardness evolution from the base metal to the weld bead for group 1, group 2 (preheated at 450 °C) and group 3 samples (annealed at 900 °C).

Figure 3 shows the wear tracks on the surface of the coupons after the pin-on-disc tests. The profile of the tracks at the weld bead and at the base metal can be seen in Figure 4.

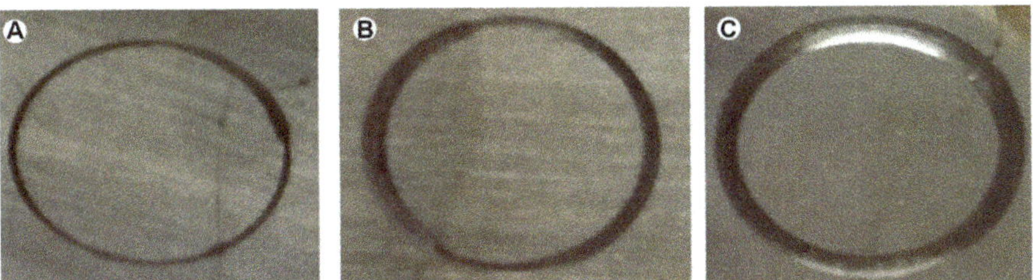

Figure 3. Wear tracks for (**A**) group 1, (**B**) group 2 (preheated at 450 °C) and (**C**) group 3 samples (annealed at 900 °C).

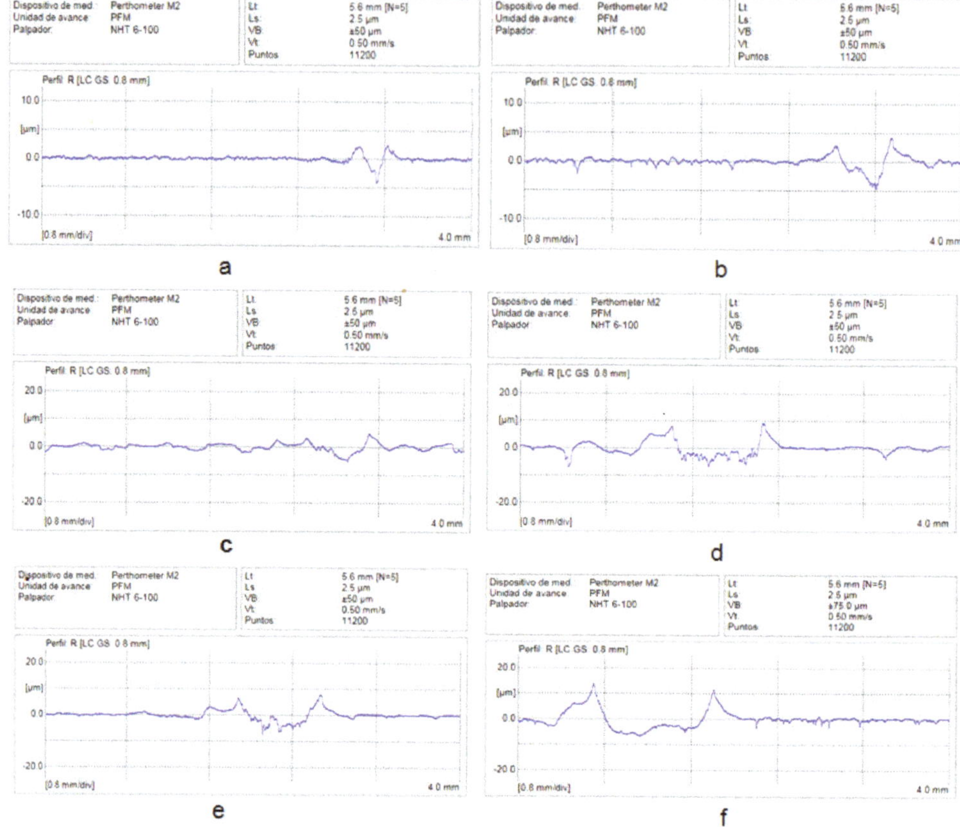

Figure 4. Wear graphs: (**a**,**b**) Without thermic treatment. (**c**,**d**) With preheating 450 °C. (**e**,**f**) Annealed at 900 °C.

4. Analysis of Results

Figure 5 shows the microstructure of the HAZ, the weld bead and the interface between them for group 1 coupons. The microstructure at the weld bead (Figure 5a) is that of a grey cast iron with compacted graphite [13]. The HAZ (Figure 5b) show large graphite nodules surrounded by ferrite in a perlitic matrix where some small precipitates of iron carbide are present. The most important changes have taken place next to the fusion line (Figure 5c), where hard ledeburite coexists along graphite nodules. The presence of such a hard microconstituent raise the hardness of the zone to 510 HV, much higher than the one measured at the weld bead or the HAZ [14], where hardness decreases rapidly by one-half. A notable hardness increment with respect to the base or the filler metal was found around the fusion line.

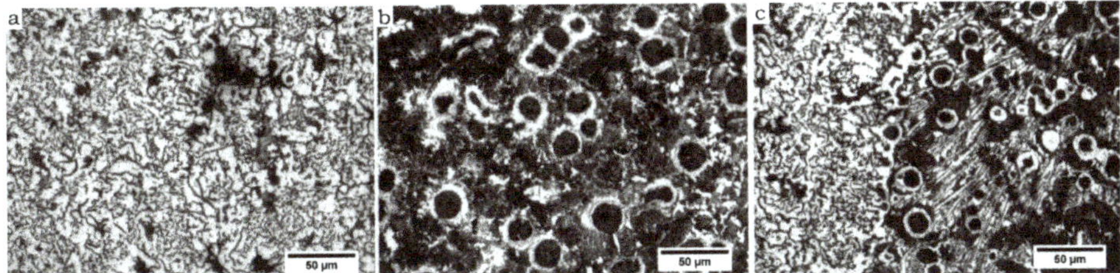

Figure 5. Micrograph of the joint without preheating nor post weld heat treatment. (**a**) Weld bead. (**b**) HAZ. (**c**) Interface.

This change in the microstructure leads to a more brittle behaviour of the joint in comparison with the base metal, even though the mechanical properties of the joint are higher than those of the filler metal, perhaps due to the effect of the part of the base metal that melts and mixes with the filler metal (dilution) to form the weld bead, changing its composition. Elongation, that goes from 14% to 9%, is the most affected parameter.

In the tensile tests the breakage took place at the weld bead (Figure 6A) and the crack faces show the intergranular aspect that can be seen in the scanning electron microscopy (SEM) image shown in Figure 7.

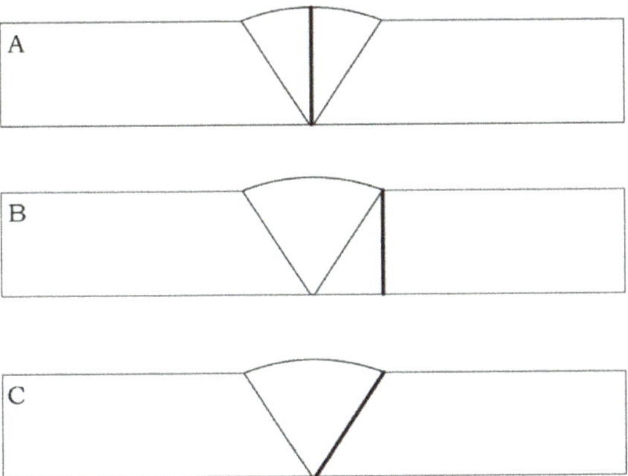

Figure 6. Location of the break in the tensile tests. (**A**) At the weld bead in group 1, (**B**) at the HAZ in group 2 and (**C**) at the fusion line in group 3.

Figure 7. SEM image of the broken surface for group 1 samples.

Figure 8 corresponds to the different microstructures found at the joint after an annealing at 900 °C for 1 h and subsequent slow cooling in the furnace. A notable difference is observed with respect to the previously described microstructure, in which no treatment had been applied.

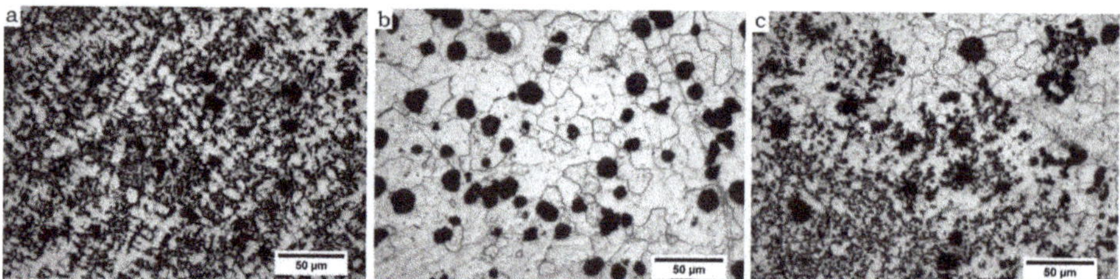

Figure 8. Micrograph of the joint after an annealing at 900 °C. (**a**) Weld bead. (**b**) HAZ. (**c**) Interface.

The most important change (Figure 8c) is the absence of ledeburite or any other rest of fragile microstructures next to the fusion line. In its place, a fast transition between the weld bead and the base material is found. The effect of the heat treatment leads at the HAZ (Figure 8b) [15] to the disappearance of the perlitic-ferritic matrix, that is replaced by a fully ferritic matrix with a hardness below 174 HV. Similar hardness values were measured at the weld bead (Figure 8a) [16]. The main difference between the microstructure of the weld bead and the HAZ is the graphite shape, compacted or vermicular at the weld bead and fully nodular at the HAZ [17].

Breakage took place at the HAZ as shown in Figure 6B. The aspect of the broken surface can be seen in Figure 9. In this case, the aspect of the fracture corresponds to a more ductile behaviour at a ferritic matrix.

Figure 9. SEM image of the broken surface for group 2 samples.

The softening of the microstructure caused by the annealing improves elongation, that now reaches a value of 12% but is also accompanied by a drop in the values of yield and ultimate strength. The new values of these variables are now very similar to the ones of the filler metal but with a higher elongation. This still represents a loss of properties for the welded plates in every aspect.

The microstructure (Figure 10) of the group 3 coupons, welded after preheating the plates at 450 °C could be described as a mixture between the two previously described. No hard structures of cementite can be observed next to the fusion line (Figure 10c), an effect of the slower cooling rate caused by having heatedthe plates at 450 °C before welding. At the HAZ (Figure 10b) graphite nodules surrounded by ferrite are observed in a perlitic matrix. The higher harness values are found at the HAZ, around 0.5 mm from the fusion line. At this zone, values near 350 HV were measured, while at the fusion line hardness was only 225 HV. This location of the most hardened material is due to the coincidental combination of maximum reached temperature and cooling rate of each point of the joint. At the weld bead hardness in maintained always below 240 HV.

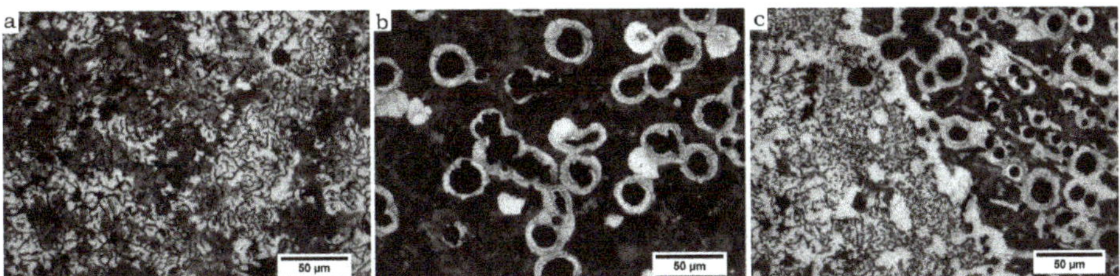

Figure 10. Micrograph of the joint preheated at 450 °C. (**a**) Weld bead. (**b**) HAZ. (**c**) Interface.

In these coupons the hardness distribution is inverted in relation to the original materials, where the higher hardness corresponds to the filler [18]. The microstructure at the weld bead show vermicular graphite in a ferritic-perlitic matrix, although some irregular graphite nodules are visible.

The achieved mechanical strength of this joints is a bit higher than the corresponding to group 2, although with a lower value of elongation.

The surface showed an aspect between ductile and brittle. The crack propagated in an intergranular way along the fusion line as indicated in Figure 6C. A SEM image of the surface is shown in Figure 11.

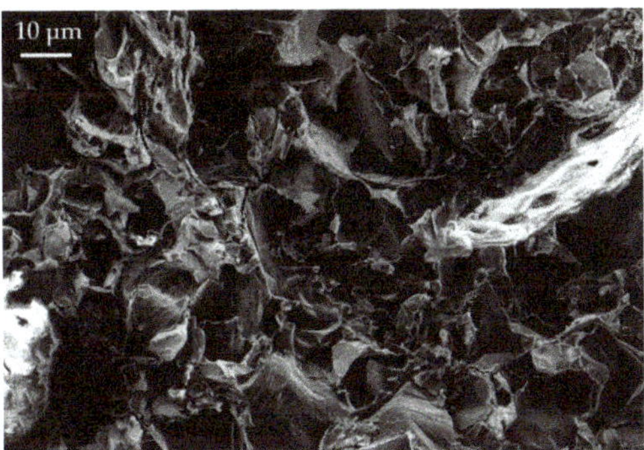

Figure 11. SEM image of the broken surface for group 3 samples.

Regarding the wear resistance of the joint, it was evaluated measuring the width and depth of the wear track both at the weld bead and at the HAZ (base material). The wear behaviour of the material near the fusion line, where ledeburite was found for group 1 coupons, was not evaluated due to the small size of this zone.

For the group 1 coupons the wear track had a width of 0.30 mm and an average depth of 2.5 microns at the HAZ and of 0.35 mm and 3 microns at the weld bead (Figure 4a,b) [19].

The width of the wear track at the HAZ for the group 2 coupons (with preheating) was 0.4 mm and the depth 4 microns. These values were 0.65 mm and 5 microns at the weld bead (Figure 4c,d).

Finally, in group 3 (annealed coupons) the wear track profile had a dimensions of 0.4 mm width and 4 microns depth at the base material and 0.9 mm width and 7 microns depth at the weld bead (Figure 4e,f).

Lower hardness values lead to higher wear and a bigger track as the resistance to abrasive wear diminishes but hardness is not the only influencing variable, as can be seen in the fact that wear is always higher at the weld bead than at the base metal even when both zones have the same hardness [20,21]. This indicates that the microstructure and the shape adopted by graphite plays a crucial role, being the vermicular graphite the most susceptible to wear.

5. Conclusions

In summary, the use of grey cast iron as filler material reduces the hardness of the fusion line with respect to the use of Ni alloys, case in which values of 700 HV are obtained [3,17,18], facilitates the dissolution of these hard microstructures with an annealing and even avoids its apparition when preheating at 450 °C, what does not happen with Ni alloys and mechanical characteristics similar to those obtained with Ni-based electrodes are reached [5,17,18]. Nevertheless, the use of grey cast iron, with lower mechanical characteristics than the base metal, limits the properties to those of the filler alloy.

The maximum loss of elongation corresponds to the weld carried out without preheating or postweld heat treatment, although it has the higher strength. These welds have the hardest microstructure with presence of ledeburite at the fusion line.

A preheating at 450 °C is enough to avoid very hard structures at the fusion line and a certain recovery of ductility, avoiding the high cost of an annealing.

The most important ductility recovery happens when the joint is subjected to an annealing treatment at 900 °C. In this case elongation reaches 12%, although the values of ultimate and yield strength reaches their minimum. With this treatment hardness also reaches its lower values in all the weld zones, below the hardness of the base material, that now has a ferritic matrix due to the ferritizing effect of the annealing.

The changes that take place in the microstructure lead to a different breakage location for each test group. At the weld bead for group 1 samples, at the HAZ for group 2 samples, when the weld was annealed and at the fusion line when the plates were preheated.

As regards the wear behaviour of the joint, the most hard structures are, as expected, also the lest affected by wear. This means the joints carried out without preheating and not annealed are the most resistant to wear and the less resistant the annealed ones, with wear track widths of 0.35 and 0.9 mm each one at the weld bead.

Another important result is that the wear track width is always wider and deeper at the weld bead than at the HAZ even when the hardness of both zones are very similar. This means the vermicular shape of the graphite at the weld bead makes this material more susceptible to wear than the graphite nodules of the base material.

Author Contributions: In this investigation, F.-J.C.-C. and M.P.-G. conceived and designed the experiments; F.-J.C.-C., F.S.-V. and V.D.-Q. performed the experiments; F.-J.C.-C., F.S.-V. and M.P.-G. analysed the data; V.D.-Q. contributed materials/analysis tools; F.-J.C.-C. and M.P.-G.wrote the paper.

Funding: This research received no external funding.

Acknowledgments: We would like to give thanks to materials technology institute (ITM) of Universitat Politècnica de València, Spain, for supporting this research.

References

1. Lacaze, J.; Sertucha, J.; Larrañaga, P.; Suárez, R. Statistical study to determine the effect of carbon, silicon, nickel and other alloying elements on the mechanical properties of as-cast ferritic ductile irons. *Rev. Metal.* **2016**, *52*. [CrossRef]
2. Suárez Sanabria, A.; Fernández Carrasquilla, J. Microestructura y propiedades mecánicas de una fundición esferoidal ferrítica en bruto de colada para su uso en piezas de grandes dimensiones. *Rev. Metal.* **2006**, *42*, 18–31.
3. Pouranvari, M. On the weldability of grey cast iron using nickel based filler metal. *Mater. Des.* **2010**, *31*, 3253–3258. [CrossRef]

4. Gouveia, R.M.; Silva, F.J.; Paiva, O.C.; de Fátima Andrade, M.; Pereira, L.A.; Moselli, P.C.; Papis, K.J. Comparing the Structure and Mechanical Properties of Welds on Ductile Cast Iron (700 MPa) under Different Heat Treatment Conditions. *Metals* **2018**, *8*, 72. [CrossRef]

5. Pascual-Guillamón, M.; Cárcel-Carrasco, J.; Pérez-Puig, M.Á.; Salas-Vicente, F. Influence of an ERNiCrMo-3 root pass on the properties of cast iron weld joints. *Rev. Metal.* **2018**, *54*, e122. [CrossRef]

6. Milosan, I. The Manufacturing of a Special Wear-resistant Cast Iron Used in Automotive Industry. *Procedia Soc. Behav. Sci.* **2014**, *109*, 610–613. [CrossRef]

7. Oksanen, V.; Valtonen, K.; Andersson, P.; Vaajoki, A.; Laukkanen, A.; Holmberg, K.; Kuokkala, V.T. Comparison of laboratory rolling–sliding wear tests with in-service wear of nodular cast iron rollers against wire ropes. *Wear* **2015**, *340–341*, 73–81. [CrossRef]

8. Medyńskia, D.; Janus, A. Effect of heat treatment parameters on abrasive wear and corrosion resistance of austenitic nodular cast iron Ni–Mn–Cu. *Arch. Civ. Mech. Eng.* **2018**, *18*, 515–521. [CrossRef]

9. Cárcel-Carrasco, F.J.; Pérez-Puig, M.A.; Pascual-Guillamón, M.; Pascual-Martínez, R. An Analysis of the Weldability of Ductile Cast Iron Using Inconel 625 for the Root Weld and Electrodes Coated in 97.6% Nickel for the Filler Welds. *Metals* **2016**, *6*, 283. [CrossRef]

10. Abboud, J.H. Microstructure and erosion characteristic of nodular cast iron surface modified by tungsten inert gas. *Mater. Des.* **2012**, *35*, 677–684. [CrossRef]

11. Sellamuthu, P.; Samuel, D.G.; Dinakaran, D.; Premkumar, V.P.; Li, Z.; Seetharaman, S. Austempered Ductile Iron (ADI): Influence of Austempering Temperature on Microstructure, Mechanical and Wear Properties and Energy Consumption. *Metals* **2018**, *8*, 53. [CrossRef]

12. Islam, M.A.; Haseeb, A.S.M.A.; Kurny, A.S.W. Study of wear of as-cast and heat-treated spheroidal graphite cast iron under dry sliding conditions. *Wear* **1995**, *188*, 61–65. [CrossRef]

13. El-Banna, E.M. Effect of preheat on welding of ductile cast iron. *Mater. Lett.* **1999**, *41*, 20–26. [CrossRef]

14. Askari-Paykani, M.; Shayan, M.; Shamanian, M. Weldability of ferritic ductile cast iron using full factorial design of experiment. *Int. J. Iron Steel Res.* **2014**, *21*, 252–263. [CrossRef]

15. Hütter, G.; Zybell, L.; Kuna, M. Micromechanical modeling of crack propagation in nodular cast iron with competing ductile and cleavage failure. *Eng. Fract. Mech.* **2015**, *147*, 388–397. [CrossRef]

16. Zhang, Y.Y.; Pang, J.C.; Shen, R.L.; Qiu, Y.; Li, S.X.; Zhang, Z.F. Investigation on tensile deformation behavior of compacted graphite iron based on cohesive damage model. *Mater. Sci. Eng.* **2018**, *713*, 260–268. [CrossRef]

17. Pascual, M.; Cembrero, J.; Salas, F.; Martínez, M.P. Analysis of the weldability of ductile iron. *Mater. Lett.* **2008**, *62*, 1359–1362. [CrossRef]

18. El-Banna, E.M.; Nageda, M.S.; El-Saadat, M.A. Study of restoration by welding of pearlitic ductile cast iron. *Mater. Lett.* **2000**, *42*, 311–320. [CrossRef]

19. Cui, J.; Chen, L. Microstructure and abrasive wear resistance of an alloyed ductile iron subjected to deep cryogenic and austempering treatments. *J. Mater. Sci. Technol.* **2017**, *33*, 1549–1554. [CrossRef]

20. Cárcel-Carrasco, J.; Pascual, M.; Pérez-Puig, M.; Segovia, F. Comparative study of TIG and SMAW root welding passes on ductile iron cast weldability. *Metalurgija* **2017**, *56*, 91–93.

21. Cárcel-Carrasco, F.J.; Pascual-Guillamón, M.; Pérez-Puig, M.A. Effects of X-rays radiation on AISI 304 stainless steel weldings with AISI 316L filler material: A study of resistance and pitting corrosion behavior. *Metals* **2016**, *6*, 102. [CrossRef]

Article

Investigation of Drilling Machinability of Compacted Graphite Iron under Dry and Minimum Quantity Lubrication (MQL)

Yang Li [1] and Wenwu Wu [2],*

1   State Key Laboratory for Manufacturing System Engineering, Xi'an Jiaotong University, Xi'an 710054, China; liyangxx@xjtu.edu.cn
2   Key Lab. of NC Machine Tools and Integrated Manufacturing Equipment of the Education Ministry & Key Lab of Manufacturing Equipment of Shaanxi Province, Xi'an University of Technology, Xi'an 710048, China
*   Correspondence: wuwen@xaut.edu.cn; Tel.: +86-1559-499-2919

Received: 4 September 2019; Accepted: 8 October 2019; Published: 11 October 2019

Abstract: Compacted graphite iron (CGI), which is used as a potential material in the auto industry, is a hard-to-machine material for the different minor elements and for the geometry of graphite with grey cast iron. The machinability of CGI in the drilling process was investigated with a 4-mm diameter fine-grain carbide twist drill under four lubrication conditions, dry (no compressed air), dry (with compressed air), MQL 5 mL/h, and MQL 20 mL/h in this paper. The maximum flank wear, types of wear, and cutting loads were studied for identifying the wear mechanism in drilling of CGI. The tool life in the four experiments of CGI drilling is 639 holes, 2969 holes, 2948 holes, and 2685 holes, respectively. The results showed that the main wear mechanism in drilling of CGI is adhesion and abrasion. Carbon, which originates from the graphite of CGI, can improve the lubrication in the drilling process by comparing with MnS in drilling grey cast iron. The thrust force and torque are more than 1000 N and 150 N*cm after 2700 holes in CGI drilling. Drilling of CGI under dry conditions (with compressed air) and MQL 5 mL/h is feasible.

Keywords: compacted graphite iron; minimum quantity lubrication (MQL); drilling machinability; dry machining

1. Introduction

Compacted graphite iron (CGI) has been widely utilized for auto parts such as engine block and head because the higher mechanical and thermal loads in the engine are required for abating pollution [1]. The engine manufactured by CGI can provide higher power, lighter weight, thinner wall thickness, and the same elastic buffer compared with the grey cast iron (GI) engine. In contrast with the aluminum engine, the CGI engine have advantages on price, duration, vibration, noise, and energy consumption. Since CGI has higher tensile strength and lower thermal conductivity than GI, the cutting loads are relatively higher while machining CGI [2]. When machining GI, the MnS layer can reduce the oxidation and diffusion during the machining processing, but gives rise to chemical wear, as well as lower tool life. By contrast, in the CGI machining process, it is very difficult to form the MnS layer because the residual sulfur in CGI has combined with magnesium in vermiculizer during the casting process [3]. Therefore, it is very important to ensure the tool life while machining CGI.

Machining of CGI has been studied in all traditional machining methods, such as turning, milling, and drilling [4–9]. Rose et al. [8] studied the effect of the content of Ti on tool wear and surface roughness in CGI turning. The results indicated that the increasing content of Ti in CGI could dramatically decrease the tool life. According to Dawson's research, the tool life in CGI turning could be reduced by 50% while the content of Ti in the CGI increased from 0.01% to 0.02% [5]. Nayyar et al. [6] investigated the effect of tool geometry on cutting load and tool life in turning CGI and found out that a smaller radius of cutting edge could induce lower cutting force and higher tool life. Abele et al. [7] utilized liquid CO_2 as lubricant in CGI turning and successfully double the productivity. Da Silva et al. [9] investigated the tool wear effect of cemented carbide coating tools in CGI milling and concluded that the flank wear of the tool was reduced with the increase in cutting speed. Accordingly, the machinability of CGI has always been a hot research topic.

The main wear mechanism of the tool in the CGI machining process is within the adhesive because the casting surface of CGI consists of the ferrite, which is easy to bond with the cutting edge of the tool. In addition, abrasive wear has been observed in the prior study [10–14]. The flank surface of the cutting tool is the main region where tool wear can be found after the machining. De Oliveira et al. [10] conducted the CGI drilling experiments, which used three tool-based geometries of the cemented carbide drill with TiAlN-coating, and discovered that the main wear mechanism of the drill is abrasive wear in these experiments. Gabaldo et al. [11] researched the tool wear mechanism and tool life in CGI milling using a cemented carbide and ceramic tool. In the finish milling, the tool made of cemented carbide had higher tool life than the ceramic tool. The effects of different commercial coating in drilling of CGI were studied by Paiva in Reference [14]. The results indicated that adhesion occurred at the cutting edge after coating breakage. In this paper, the wear mechanisms and cutting loads in drilling of CGI using three different lubrication conditions were also investigated.

Walter, which is a famous tool maker company, suggested that no lubricant should be used in high-speed machining of cast iron because the lubricant can reduce the tool life. Nayyar et al. [15] studied the machinability of GI, CGI, and ductile iron. The tool life was found to decrease when using a lubricant in the machining of GI. Yet, a reduction in thrust force and higher tool life was found during the machining of CGI and ductile iron at 200 m/min cutting speed with a coated tool. Heck et al. [16] investigated the effect of the MnS layer in GI and CGI turning using several inspection methods and concluded that multiple insert tools is a good method to obtain high productivity in machining CGI. Quinto [17] surveyed the chemical vapor deposition(CVD) and Physical Vapor Deposition (PVD) hard-coating methods and considered that the optimal combination of tool material, geometry, and coating was an important way to realize the high productivity machining of new difficult-to-cut material. Kuzu et al. [18,19] presented a new thermal modeling method to predict the drill temperature of CGI drilling. Wu et al. [20] investigated the feasibility of CGI drilling under dry and the MQL condition through the experiment. Da Mota studied the wear mechanisms of tapping CGI and found that the adhesive and abrasive mechanisms are the main modes [21]. Compared with the traditional machining process, high spindle speed and feed rate are used in a high throughput machining process to obtain high machining efficiency and reduce the part cost. The wear mechanism of the drill bit in high throughput CGI drilling is studied in this paper.

High throughput machining of CGI still remains a problem that needs further research. In this paper, the effects of three lubrication conditions on the tool wear and cutting load in drilling of CGI are studied. Experimental setup is introduced in Section 2. The testing results of tool wear and cutting load with an increment of holes are introduced and discussed in Section 3. Lastly, conclusions are presented in Section 4.

2. Experimental Setup

In the drilling experiments, the compacted graphite iron (CGI) workpiece plates (270-mm length, 206-mm width, and 32-mm thickness) were used. In order to optimize castability and mechanical properties of CGI, the material chemical composition has more than 80% pearlite, contains maximum 0.015% titanium, maximum 0.10% chromium, and maximum 0.40% manganese. The mechanical properties of CGI are 420 MPa in ultimate tensile stress, 140 GPa elastic modulus, and 210–265 BHN in hardness. For each CGI plate, the casting surface was machined at 1 mm for obtaining the flat surface and eliminating the heterogeneous part of the casting material. In the experiments, 25-mm depth holes were drilled and the tool wear was measured. Figure 1 shows the dimension and holes layout of the CGI plate, which is used in the drilling experiment.

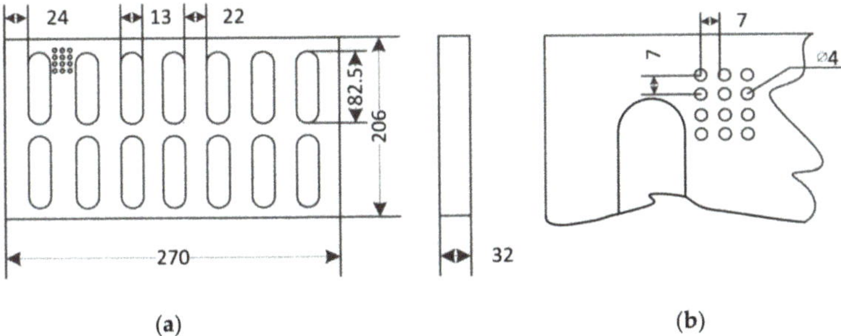

Figure 1. Dimension and hole layout of the CGI plate (Unit: mm). (**a**) The dimension of the CGI plates, and (**b**) holes layout.

The drill is a fine-grain carbide twist drill (B255A04000YPC KCK10, Kennametal, PA, USA). It has multi-layers (AlCrN base layer, TiAlN/AlCrTiN middle layers, and AlCrN outer layer) obtained by the drill supplier Kennametal. The diameter of the drill is 4 mm in diameter and the point angle is 135°. This drill has special designs with three margins, two through-the-drill 0.7-mm diameter holes, and an S-shape chisel. The top and side views of the drill are shown in Figure 2.

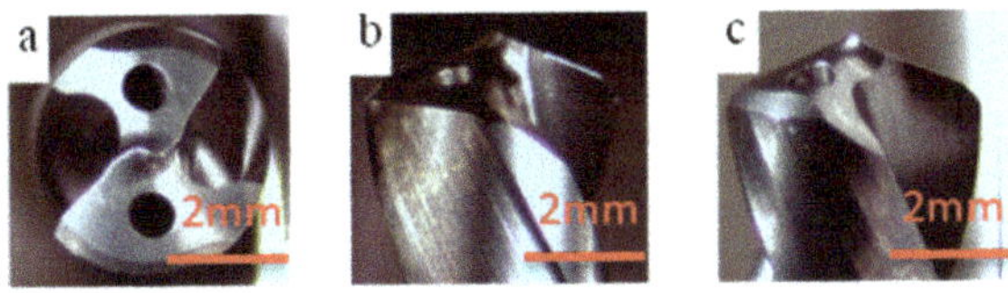

Figure 2. The drill applied in the experiments: (**a**) Top view, (**b**) side view 1, and (**c**) side view 2. Reproduced from [20], with permission from Taylor & Francis.

The drilling experiments were carried on a vertical machining center (Model 4020 by Fadal, Troy, MI, USA) shown in Figure 3. The single channel MQL system was mounted on the machine center made by UNIST, USA. The MQL system utilizes a high-speed air stream of compressed air to atomize the lubricant and blow it to the cutting area for cooling and lubrication. The compressed air (690 kPa) was applied in dry and MQL conditions. In the MQL condition, the Coolube 2210EP by UNIST was used for the cutting fluid.

In order to compare the drilling machinability under different lubrication conditions, four experiments, marked as Experiment I, II, III, and IV, are conducted (Table 1). In Experiment I–IV, dry, air, 5 mL/h MQL, and 20 mL/h MQL, were applied as lubrication, respectively. For comparing the tool wear mechanism, grey cast iron is drilled under a 5 mL/h MQL lubrication condition and marked as Experiment V. In the experiments with MQL, the cutting fluid 2210 EP (Unist Inc., Grand Rapids, MA, USA) was also applied. All of the experiments were conducted at a feed rate of 0.2 mm/rev and a constant cutting speed of 100 m/min. The spindle speed is 7961 rpm (the maximum spindle speed is 15,000 rpm). The number of holes before the drill breakage represents the drill life in Table 1.

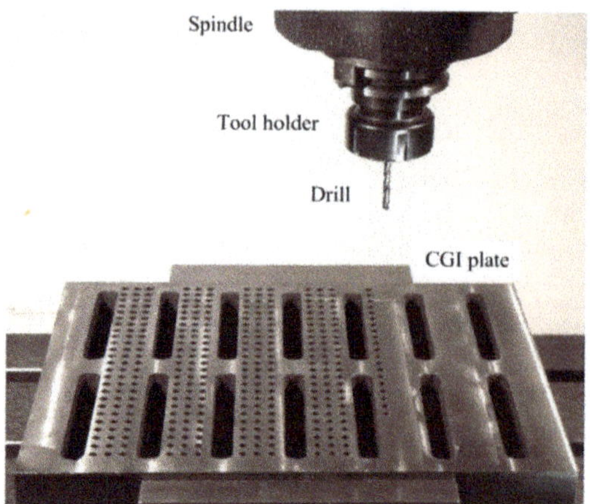

Figure 3. Drilling experiment setup.

Table 1. Five drilling experiments.

Exp.	Workpiece Material	Lubrication Condition	Drill Life
I	CGI	Dry (no compressed air)	639
II	CGI	Dry (with compressed air)	2969
III	CGI	MQL, 5 mL/h	2948
IV	CGI	MQL, 20 mL/h	2685
V	GI	MQL, 5 mL/h	>3000

The tool wear was measured after every 150 holes until 900 holes were drilled. After this, the tool wear was measured after every 300 holes. The tool wears were magnified 100 times and observed by using an optical microscope. The Philips XL 30 with an integrated energy dispersive X-ray spectroscopy (EDS) system was applied to obtain the scanning electron microscope (SEM) picture and analyze the elements of the worn drill.

In order to obtain the cutting load during the drilling process, a two-components dynamometer (type 9271A, Kistler, Switzerland), which can measure the torque and thrust force, was adopted. The amplifier and the software developed by LabVIEW were applied to process and store the testing data. When measuring the cutting load in different holes, the small CGI plate was cut out from the CGI plate (shown in Figure 3) and mounted to the dyno. In every 300 holes, the torque and thrust force were measured for each experiment.

3. Results and Discussion

3.1. Tool Wear

3.1.1. Maximum Flank Wear

Since the flank surface is directly related to the tool wear and tool life, the maximum flank wear of the drilling was investigated in this paper. Figure 4 illustrates the maximum flank wear in Experiment I to V.

From Figure 4, we know that the tool wear in Experiment I increases linearly with the number of holes before the drill is broken. In Experiment II and III, the process of tool wear can be divided into three phases: break in the period (0–900 holes), steady state wear region (90–2100 holes), and failure region (2101-drill broken). The tool life in the first four experiments is 639 holes, 2969 holes, 2948 holes, and 2685 holes, respectively. The tool life in Experiment I is the shortest. This is because, in Experiment

I, there is no lubrication and the maximum flank wear is higher than those in Experiments II–IV. Furthermore, in the first 1800 holes in Experiment II and Experiment III, the flank wear of the latter is slightly lower than the former. After that, the differences of flank wears in these two experiments grow bigger. These differences indicate that, when applying the MQL (5 mL/h), the lubrication condition has been improved and the wear has been reduced. However, if the spray speed of MQL changes to 20 mL/h, the friction between the tool and the workpiece cannot be improved. As shown in Figure 4, the maximum flank wear and wear rate in Experiment IV is higher than that in Experiments II and III. The mixture of oil and dust chips in Experiment IV is difficult to eject, which causes higher tool wear. In conclusion, the good chip evacuation and the lubrication of the oil could lead to the lowest tool wear (Experiment III).

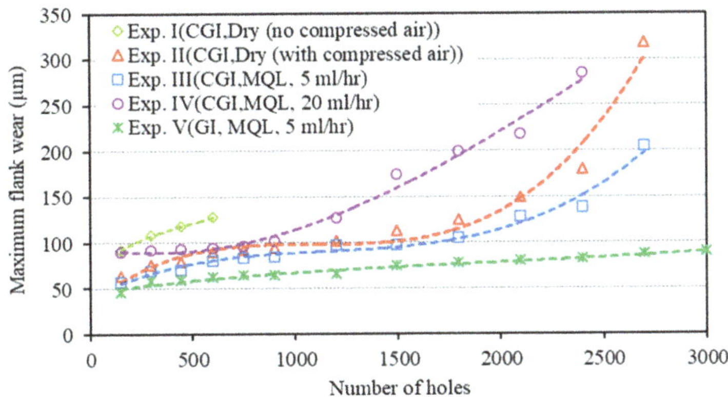

Figure 4. The maximum flank wears in five experiments. Reproduced from [20], with permission from Taylor & Francis.

In addition, after measuring the cutting length in those experiments, the cutting length in Experiment I was found to be the shortest (16 m) when comparing to the ones obtained from Experiments II and III (74.2 m and 73.7 m). The results of cutting length are inconsistent with the results of tool wear. There are two reasons: the lowest flank wear in Experiment III is caused by the lubrication of the oil and the longest cutting length in Experiment II is brought about by the stability of the compressed air by giving rise to continuous chip evacuation. Compared to the experimental results of compressed air, the tool life under the MQL condition is lower because of the instability of the MQL compressed air. The instability is caused by the power consumption for generating the oil droplet.

In Experiment V, the GI plate is drilled. It can be observed from Figure 4 that the tool wear and wear rate in drilling GI is lower than in drilling CGI. After 3000 holes, the maximum flank wear in Experiment I is still small (less than 100 micron). If the tool wear in drilling GI has a linear relationship with the number of holes, another 3000 holes could be drilled before the tool wear reaches 150 micron. GI has good machinability in the drilling process when compared with CGI.

3.1.2. The Worn Drills

Figures 5–7 show the flank wear, crater wear, and margin wear of the worn drills in Experiments I–III at the point before being broken.

Figure 5 presents the flank wears of three worn drills at five points in Experiments I–III, respectively. Flank wear occurs along the relief surface of the cutting edges and it is primarily caused by abrasive wear. The scratch can also be observed near point D in Experiment II and III as a result of abrasive wear. From Figure 5, it can be concluded that tool wear at five points increased with the distance from the center of the drill. Points D and E are far from the center of the drill and have the largest tool wear. The tool wears at points A and B are smaller than that at points D and E. In addition, they are almost the same. The wear rate at point C, which is situated at the chisel edge, is the lowest. The scratch can also be observed near point D in Experiment II and III as a result of abrasive wear.

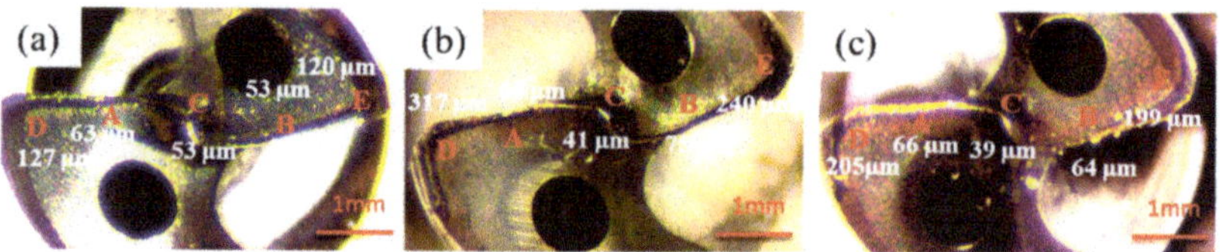

Figure 5. The flank wears of three worn drills: (**a**) Experiment I (600th hole), (**b**) Experiment II (2700th hole), and (**c**) Experiment III (2700th hole). Reproduced from [20], with permission from Taylor & Francis.

Figure 6 illustrates the crater wears of three worn drills. The crater wears occur on the flute face near the cutting edges. The widths of crater wears in Experiment I are 286 um and 218 um, which are more than in Experiment II and III. It can be observed from Experiment II that there are two regions of crater wear due to the friction of the compressed air coming out of the through-the-drill hole close to the cutting edge. The width of the first region is about 100 microns and the width of the second region is more than 500 microns. By contrast, there is only one reign of the crater wear in Experiment III and the width is approximately 130 microns. It is wider than the first region and thinner than the second region in Experiment II. The oil, which helps reduce the friction between the drill and workpiece, can be clearly observed on both flutes of the drill.

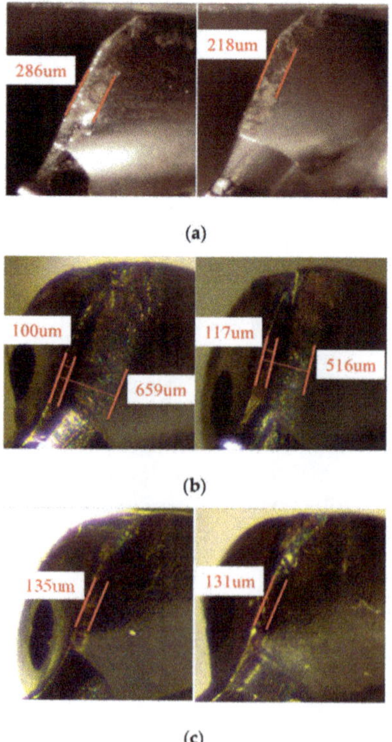

Figure 6. The crater wears of three worn drills: (**a**) Experiment I (600th hole), (**b**) Experiment II (2700th hole), and (**c**) Experiment III (2700th hole).

The margin wears of three worn drills, which occur on the outer corner of the cutting edges, are demonstrated in Figure 7. The area of the margin is used to evaluate the level of the margin wear. At the margin region, the peripheral speed is the highest. Increasing the cutting speed could increase the cutting temperature, which further causes the hardness of the drill material to drop. As a result, the main wears are abrasion, thermal softening, and diffusion in the margin region. Thermal softening not only leads to the increased abrasive wear, but also results in plastic deformation of the cutting edges. The latter would further aggravate the drill abrasion. After drilling 600 holes in Experiment I, the wear area of the two margins are 6.5×10^4 and 6.58×10^4 um^2, respectively. Apart from the heat effect zone in Experiment I, the chipping and abrasion can be clearly seen in the margin of the drill in both Experiment II and Experiment III. Compared with the margin wears in the other two experiments, the wear areas of two margins in Experiment II (3.55×10^5 and 3.04×10^5 um^2) are larger.

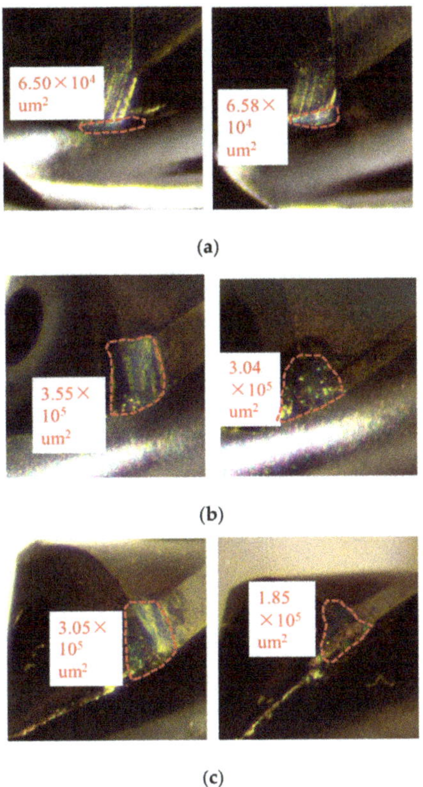

Figure 7. The margin wears of three worn drills: (**a**) Experiment I (600th hole), (**b**) Experiment II (2700th hole), and (**c**) Experiment III (2700th hole).

In conclusion, oil produced by the MQL system can be spread to the cutting drill, but only a little oil reaches the margin area. Therefore, more oil in Experiment III cannot reduce the friction between the margin and the wall of the hole. However, the pressure of the compressed air decreases due to producing more oil in Experiment III, which results in reducing the effect of the chip evacuation. Thus, the larger torque and force may occur in Experiment III. This can explain why there is lower tool wear and shorter tool life in Experiment III than in Experiment II.

3.2. Scanning Electron Microscope (SEM) and Energy Dispersive Spectrometer (EDS) Results

The rake face of the drill contacts the chip in the drilling process. The diffusion and adhesion can easily come up in these zones due to the high contact pressure. The SEM picture of the new drill is shown in Figure 8. In the present work, the worn drills after 900 holes were selected for SEM and for the analysis of the elements at the cutting edge using EDS. The rake face SEM pictures of the worn drill in Experiment II is shown in Figure 9.

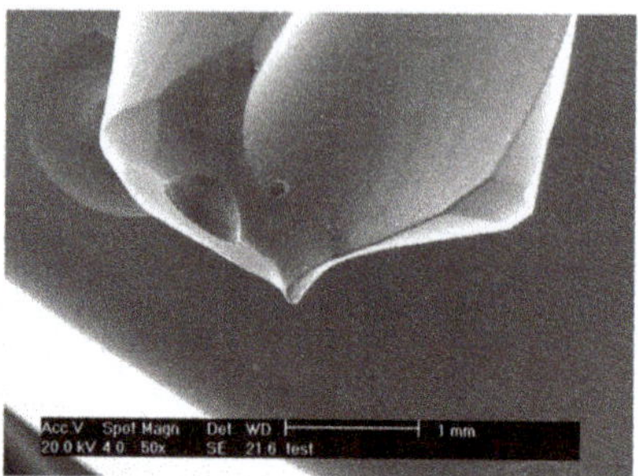

Figure 8. The SEM of one new drill.

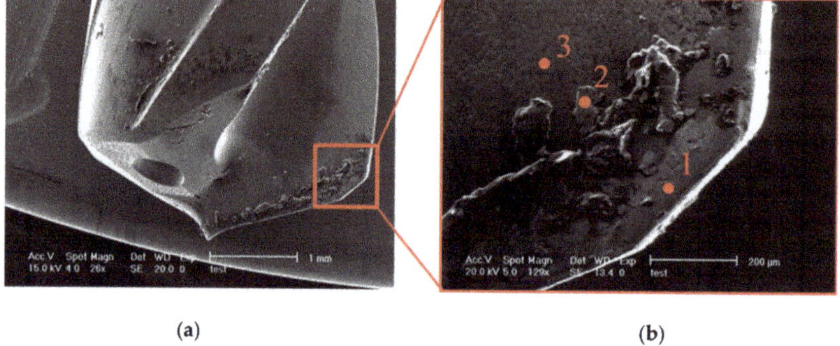

(a)   (b)

Figure 9. The SEM pictures for the worn drill after 900 holes in Experiment II: (**a**) SEM picture, and (**b**) SEM picture for EDS.

When compared with the new drill, it can be observed that some layered films are put on the rake face. Since CGI is a relatively brittle cast iron and it is very difficult to form this kind of layered films, such layered films should be the broken coating layers. Three points in different regions of the film are selected and detected by EDS to identify its element. Figure 10 presents the results of EDS at point 1, 2, and 3, respectively.

From the EDS spectra above, the integrated coating can be found in this zone. Comparing Figure 10a with Figure 10c, it can be inferred that the elements at point 1 and point 3 are similar. At the two points, a high concentration of nitrogen and carbon can be detected. Carbon is the major element of graphite, which originates from the workpiece material. Nitrogen is the main element of the coating layer. Oxygen and iron are also the key chemical elements, according to the figure. Iron is the main element of the workpiece material. It means that the adhesion occurs in the drilling process. Aluminum and chromium are also detected at these two points. They may come from the coating material. As shown in Figure 10b, the elements of the middle coating layers, such as titanium, aluminum, chromium, and nitrogen, are observed at point 2. Meanwhile, the major elements of the workpiece material (iron, carbon, and silicon) are also detected according to the results of EDS. There is an adhesion between the drill and workpiece at this point.

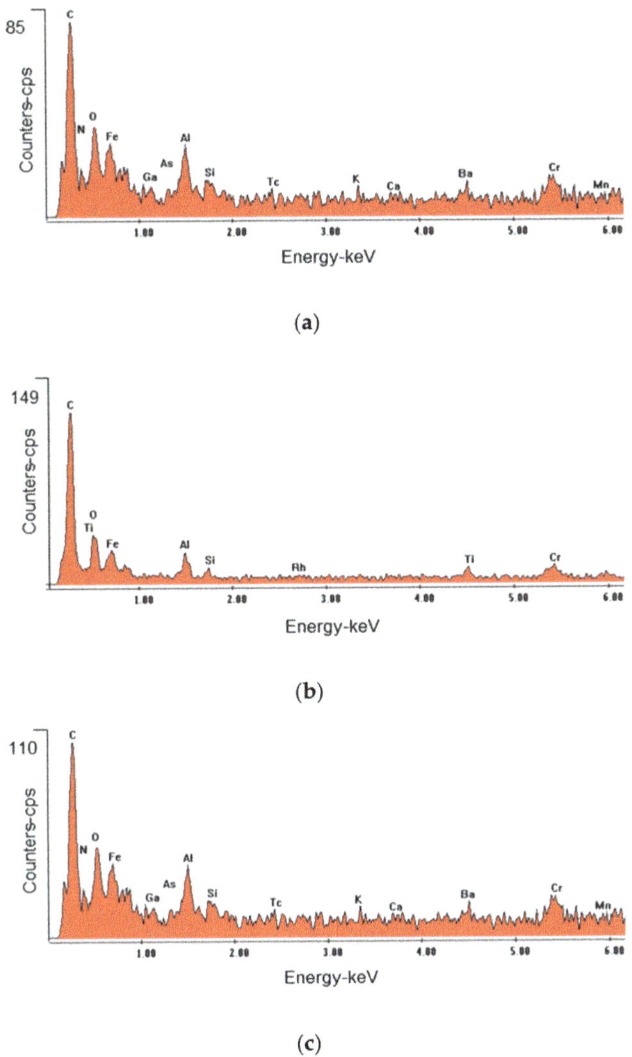

Figure 10. The EDS results at three points: (**a**) Point 1, (**b**) point 2, and (**c**) point 3.

In Section 2, the multilayers coating in this drill have been introduced in detail. They are the AlCrN base layer, the TiAlN/AlCrTiN middle layers, and the AlCrN outer layer. Comparing the results of EDS at point 1, 2, and 3 with the elements of the base layer, middle layers, and outer layer, it can be concluded that the layered film at point 2 is the middle layer (TiAlN/AlCrTiN). The coating at point 3 has a good reservation and is the outer layer. Point 1 and point 3 have basically the same elements. However, the visible damage of the coating can be easily observed at point 1. Therefore, the coating at point 1 should be the base layer (AlCrN). In addition, C and Fe can be detected at all three points and the content of carbon is clearly more than iron. Carbon in the films can be utilized as lubricant in the drilling process.

3.3. Torque and Force

3.3.1. Error Line of Torque and Thrust Force

Figure 11 shows the error line results of torque and thrust force during the drilling process in Experiments I–III.

Due to the higher wear rate, the torque and thrust force in Experiment I are higher than that in Experiments II and III. The thrust forces in Experiments II and III are exponentially increased with the number of holes and are basically similar. In the break in stage (0–900 holes), the changes of thrust force amplitude are bigger. The thrust force in the steady state wear stage (900–2100 holes) is stable. In the failure region (2101–tool break), the changes of thrust force amplitude are slightly increased when compared with the front stage. The torques tested in those three experiments have a strong relationship with the results of tool wear. Similar to the thrust force, the torque in Experiment I have the largest amplitude changes. The changes of the torques in Experiment II and III are stable. The changing trend and average values of the torque in these two experiments are almost the same in the whole drilling process. At the 2700th hole, the lower thrust force can be found when the tool wear is lower. However, the maximum torque at this hole in Experiment III is larger than that in Experiment II.

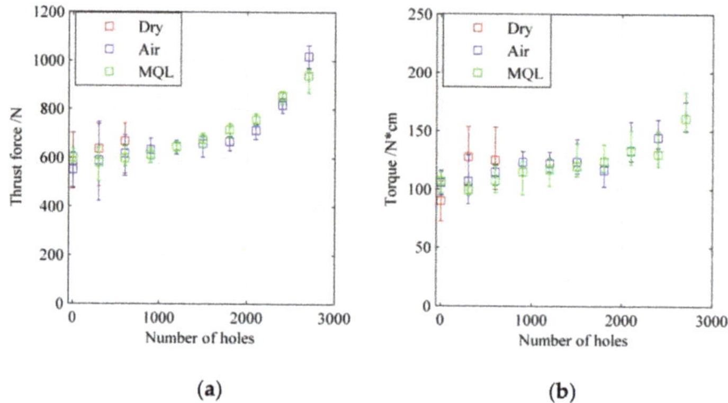

Figure 11. The error line results of cutting loads in three Experiments: (**a**) Thrust force and (**b**) torque.

3.3.2. Cutting Loads

The changes of cutting loads with the number of holes in Experiments I, II, and III are investigated. The results are shown in Figure 12.

Being affected by the chips in the holes, the cutting loads are unstable in Experiment I, as shown in Figure 12a,b. With the increase of holes, the tool wear is getting worse and worse. The torque and the thrust force are increased as well. After drilling 600 holes, the fluctuations of the lines are reduced, which means the thrust force becomes more stable. In addition, the thrust force at a specific hole does not change greatly with the drilling depth. However, by contrast, it is clear that the torque is changing (increasing at the beginning and decreasing at 600 holes) when the tool is drilling through a hole.

The thrust force in Experiment II, as shown Figure 12c, is lower and more stable than that in Experiment I. This is because, in Experiment II, the chip ejection caused by the compressed air, which comes out from the through-the-drill holes, is very effective. Similarly, as the number of holes increases, the torque and thrust force are both increased due to the increment of tool wear. At the beginning of the drilling experiment, the thrust force and torque are about 550 N and 100 N*cm. At the 2700th hole, the thrust force and torque are more than 1000 N and 150 N*cm. From Figure 12d, it can be seen that the torque at the 1st and 900th holes have slightly increased with the depth of holes. It can be explained that the friction between the margin of the drill and the wall of the hole are increased with the depth of the hole. However, the decrement of torque with the drilling depth can be observed at the 2700th hole, which may be caused by the burr formation for the large flank wear.

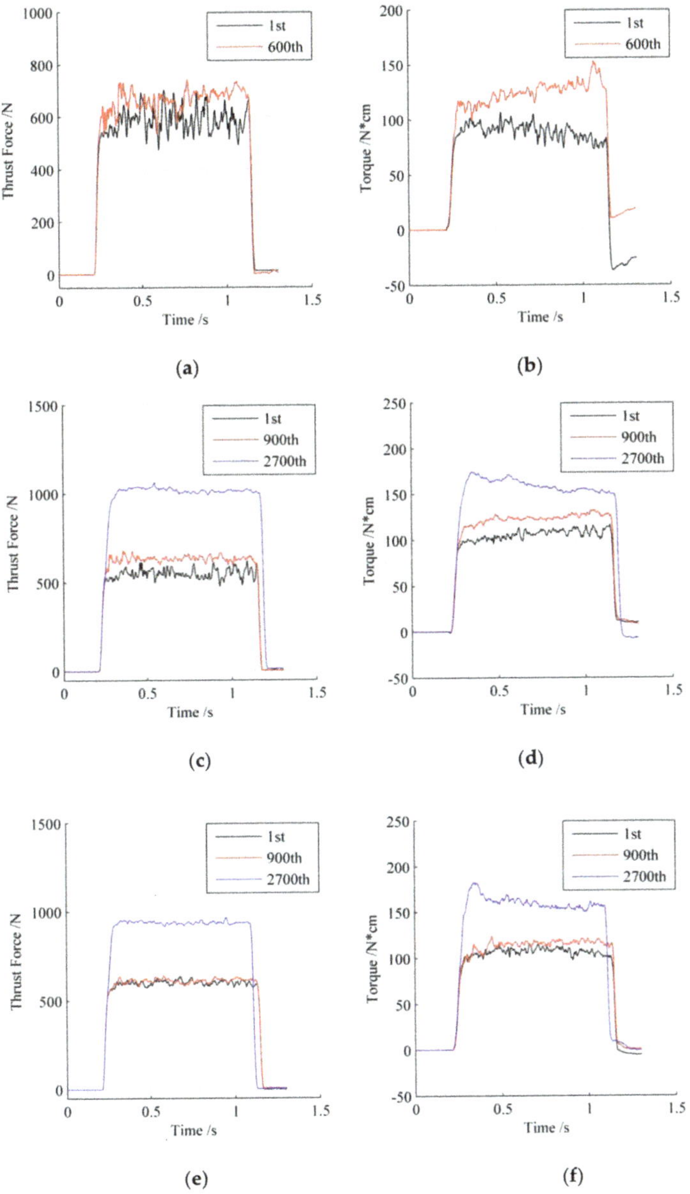

Figure 12. The cutting loads of three Experiments at different holes: (**a**) Thrust forces in Experiment I, (**b**) torques in Experiment I, (**c**) thrust forces in Experiment II, (**d**) and torques in Experiment II. (**e**) Thrust forces in Experiment III, and (**f**) torques in Experiment III. Reproduced from [20], with permission from Taylor & Francis.

Figure 12e,f present the torque and thrust force in Experiment III. Both torque and thrust force in Experiment III are lower and more stable than that in Experiment II. This is because the oil in

Experiment III can improve the lubrication between the drill and workpiece material, which further leads to lower tool wear. However, the torque at the start stage of the 2700th hole (180 N*cm) is higher than that in Experiment II (170 N*cm). Therefore, the tool life in Experiment III is shorter than that in Experiment II. After the 2700th hole, the torque in Experiment III falls quickly and becomes lower than that in Experiment II. In conclusion, because of the lubrication of oil in Experiment III, the lower and more stable cutting loads can be obtained while the hole is drilling.

4. Conclusions

The drilling machinability of CGI under different lubrication conditions are investigated in this paper. The maximum flank wear and tool wear at five points along the cutting edge are measured and investigated at first. The wear mechanism is then studied based on the SEM and EDS results. Lastly, the cutting loads during the drilling process are measured. Several conclusions for drilling CGI are summarized below.

1. Drilling of CGI under dry conditions (with compressed air) and MQL 5 mL/h is feasible. Drilling of CGI with compressed air leads to longer tool life and lower tool wear than under a dry condition, because the compressed air strongly improves the ability of chip evacuation. Drilling of CGI with MQL (5 mL/h) has the smallest tool wear due to the lubrication of oil. The small amount of oil under the MQL condition (5 mL/h) can improve the lubrication of the drilling procedure, which results in the lower tool wear and cutting loads. However, the large amount of oil (20 mL/h) can mix with the dust chip to slow the movement of chips.

2. Adhesion and abrasion wear are the main mechanisms of drilling CGI. Severe abrasion wear when drilling CGI under a dry condition occurs due to the hard dust chip, which is very difficult to eject through good geometry of a drill. Under compressed air, the chip ejection was greatly improved. Therefore, the adhesion becomes the vital reason for the tool wear.

3. Carbon, which originates from the graphite of CGI, can improve the lubrication in the drilling process when compared with MnS in drilling grey cast iron. This makes the dry drilling of CGI with compressed air feasible.

4. The multi-layers coating of the drill is useful for drilling CGI. This kind of coating is effective to prevent the friction between tool material and workpiece material. After the breakage of the coating, the tool wear rate is increased rapidly and the tool runs into the failure stage.

Author Contributions: Conceptualization, Y.L. and W.W. Methodology, W.W. Investigation, W.W. Data curation, W.W. Writing—original draft preparation, Y.L. Writing—review and editing, W.W.

Funding: The National Science Foundation of China (Grant No. 51705402), Natural Science Basic Research Plan in Shaanxi Province of China (Program No. 2019JM-148), and National Science and Technology Major Project of China (No. 2017ZX04013001) funded this research.

Acknowledgments: We thank the Ford Motor Company, UNIST, and Kennametal for their support.

Conflicts of Interest: The authors declare no conflict of interest.

References

1. Dawson, S.; Schroeder, T. Practical applications for compacted graphite iron. *AFS Trans.* **2004**, *112*, 1–9.
2. Gastel, M.; Konetschny, C.; Reuter, U.; Fasel, C.; Schulz, H.; Riedel, R.; Ortner, H.M. Investigation of the wear mechanism of cubic boron nitride tools used for the machining of compacted graphite iron and grey cast iron. *Int. J. Refract. Met. Hard* **2000**, *18*, 287–296. [CrossRef]
3. Evans, R.; Hoogendoorn, F.; Platt, E. Lubrication & Machining of Compacted Graphite Iron. *AFS Trans.* **2001**, 1–8.
4. Kuzu, A.T.; Bijanzad, A.; Bakkal, M. Experimental Investigations of Machinability in the Turning of Compacted Graphite Iron using Minimum Quantity Lubrication. *Mach. Sci. Technol.* **2015**, *19*, 559–576. [CrossRef]
5. Dawson, S.; Hollinger, I.; Robbins, M.; Daeth, J.; Reuter, U.; Schulz, H. The effect of metallurgical variables on the machinability of compacted graphite iron. *SAE Trans.* **2001**, *110*, 334–352.

6. Nayyar, V.; Alam, M.Z.; Kaminski, J.K.; Kinnander, A.; Nyborg, L. An Experimental Investigation of the Influence of Cutting Edge Geometry on the Machinability of Compacted Graphite Iron. *Int. J. Mech. Mater. Eng.* **2013**, *3*, 1–25. [CrossRef]

7. Abele, E.; Burkhard, S. Using PCD for machining CGI with a CO2 coolant system. *Prod. Eng.* **2008**, *2*, 165–169. [CrossRef]

8. Rosa, S.D.; Diniz, A.E.; Andrade, C.L.; Guesser, W.L. Analysis of tool wear, surface roughness and cutting power in the turning process of compact graphite irons with different titanium content. *J. Braz. Soc. Mech. Sci.* **2010**, *32*, 234–240. [CrossRef]

9. Da Silva, M.B.; Naves, V.T.; De Melo, J.D.; de Andrade, C.L.F.; Guesser, W.L. Analysis of wear of cemented carbide cutting tools during milling operation of gray iron and compacted graphite iron. *WEAR* **2011**, *271*, 2426–2432. [CrossRef]

10. De Oliveira, V.V.; Beltrao, P.A.; Pintaude, G. Effect of tool geometry on the wear of cemented carbide coated with TiAlN during drilling of compacted graphite iron. *WEAR* **2011**, *271*, 2561–2569. [CrossRef]

11. Gabaldo, S.; Diniz, A.E.; Andrade, C.L.; Guesser, W.L. Performance of carbide and ceramic tools in the milling of compact graphite iron—CGI. *J. Braz. Soc. Mech. Sci.* **2010**, *32*, 511–517. [CrossRef]

12. Alves, S.M.; Schroeter, R.B.; Bossardi, J.C.; de Andrade, C.L.F. Influence of EP additive on tool wear in drilling of compacted graphite iron. *J. Braz. Soc. Mech. Sci.* **2011**, *33*, 197–202. [CrossRef]

13. Mocellin, F.; Melleras, E.; Guesser, W.L.; Boehs, L. Study of the machinability of compacted graphite iron for drilling process. *J. Braz. Soc. Mech. Sci.* **2004**, *26*, 22–27. [CrossRef]

14. Paiva, J.M.; Amorim, F.L.; Soares, P.; Torres, R.D. Evaluation of Hard Coating Performance in Drilling Compacted Graphite Iron (CGI). *J. Mater. Eng. Perform.* **2013**, *22*, 3155–3160. [CrossRef]

15. Nayyar, V.; Kaminski, J.K.; Kinnander, A.; Nyborg, L. An Experimental Investigation of Machinability of Graphitic Cast Iron Grades; Flake, Compacted and Spheroidal Graphite Iron in Continuous Machining Operations. *Procedia Cirp* **2012**, *1*, 488–493. [CrossRef]

16. Heck, M.; Ortner, H.M.; Flege, S.; Reuter, U.; Ensinger, W. Analytical investigations concerning the wear behaviour of cutting tools used for the machining of compacted graphite iron and grey cast iron. *Int. J. Refract. Met. Hard* **2008**, *26*, 197–206. [CrossRef]

17. Quinto, D.T. Twenty-five years of PVD coatings at the cutting edge. *SVC Bull.* **2007**, 17–22.

18. Kuzu, A.T.; Berenji, K.R.; Bakkal, M. Thermal and force modeling of CGI drilling. *Int. J. Adv. Manuf. Technol.* **2016**, *82*, 1649–1662. [CrossRef]

19. Kuzu, A.T.; Berenji, K.R.; Ekim, B.C.; Bakkal, M. The thermal modeling of deep-hole drilling process under MQL condition. *J. Manuf. Process.* **2017**, *29*, 194–203. [CrossRef]

20. Wu, W.; Kuzu, A.T.; Stephenson, D.A.; Hong, J.; Bakkal, M.; Shih, A. Dry and minimum quantity lubrication high-throughput drilling of compacted graphite iron. *Mach. Sci. Technol.* **2018**, *22*, 652–670. [CrossRef]

21. Da Mota, P.R.; Reis, A.M.; Machado, A.R.; Ezugwu, E.O.; da Silva, M.B. Tool wear when tapping operation of compacted graphite iron. *Proc. Inst. Mech. Eng. B J. Eng.* **2013**, *227*, 1704–1713. [CrossRef]

Article

Niobium Additions to a 15%Cr–3%C White Iron and Its Effects on the Microstructure and on Abrasive Wear Behavior

Arnoldo Bedolla-Jacuinde [1,*], Francisco Guerra [1], Ignacio Mejia [1] and Uzzi Vera [2]

[1]   Foundry Department, Metallurgical Research institute, Michoacan University of Saint Nicholas of Hidalgo, PC 58000 Morelia, Michoacán, Mexico; fvguerra@umich.mx (F.G.); imejia@umich.mx (I.M.)
[2]   Faculty of Mechanical Engineering, Autonomous University of the State of Hidalgo, PC 43990 Cd Sahagun, Hidalgo, Mexico; uzzi610@gmail.com
*   Correspondence: abedollj@umich.mx; Tel.: +52-443-206-0952

Received: 10 November 2019; Accepted: 2 December 2019; Published: 7 December 2019

Abstract: From the present study, niobium additions of 1.79% and 3.98% were added to a 15% Cr–3% C white iron, and their effects on the microstructure, hardness and abrasive wear were analyzed. The experimental irons were melted in an open induction furnace and cast into sand molds to obtain bars of 45 mm diameter. The alloys were characterized by optical and electron microscopy, and X-ray diffraction. Bulk hardness was measured in the as-cast conditions and after a destabilization heat treatment at 900 °C for 30 min. Abrasive wear resistance tests were undertaken for the different irons according to the ASTM G65 standard in both as-cast and heat-treated conditions under three loads (58, 75 and 93 N). The results show that niobium additions caused a decrease in the carbon content in the alloy and that some carbon is also consumed by forming niobium carbides at the beginning of the solidification process; thus decreasing the eutectic M_7C_3 carbide volume fraction (CVF) from 30% for the base iron to 24% for the iron with 3.98% Nb. However, the overall carbide content was constant at 30%; bulk hardness changed from 48 to 55 hardness Rockwell C (HRC) and the wear resistance was found to have an interesting behavior. At the lowest load, wear resistance for the base iron was 50% lower than that for the 3.98% Nb iron, which is attributed to the presence of hard NbC. However, at the highest load, the wear behavior was quite similar for all the irons, and it was attributed to a severe carbide cracking phenomenon, particularly in the as-cast alloys. After the destabilization heat treatment, the wear resistance was higher for the 3.98% Nb iron at any load; however, at the highest load, not much difference in wear resistance was observed. Such a behavior is discussed in terms of the carbide volume fraction (CVF), the amount of niobium carbides, the amount of martensite/austenite in matrix and the amount of secondary carbides precipitated during the destabilization heat treatment.

Keywords: cast iron; high-chromium; abrasive wear; niobium alloying

---

1. Introduction

High-chromium white irons are widely used in the mineral processing industry due to their excellent wear behavior under abrasive conditions. A microstructure consisting of a network of hard eutectic M_7C_3 carbides in a mainly austenitic (as-cast) or martensitic (after a heat treatment) matrix makes these irons very suitable for severe wear applications [1–4]. To improve abrasive behavior in the as-cast iron, an increase in carbon and chromium would increase the carbide volume fraction. Other way to improve wear behavior is to apply a destabilization heat treatment to precipitate secondary carbides and to transform the austenitic matrix into a martensitic one [5–10]. However, these two actions would decrease fracture toughness in the alloy; which is important, particularly when abrasive wear is accompanied by repetitive impact. Under this basis, big efforts have been made to improve wear behavior without affecting considerably fracture toughness in the as-cast conditions. A strategy for this has been to reinforce the as-cast austenitic matrix with harder primary carbides such as niobium, vanadium or titanium carbides. The use of titanium is commonly limited to amounts lower than 2% in these alloys due to difficulty of adding higher amounts in an open furnace during the alloy making [11–13]. Additions of more than 2% titanium would only be possible by using a vacuum induction furnace. To get primary vanadium carbides, at least vanadium amounts of 5% are necessary, other way vanadium only reinforces the M_7C_3 carbide [14–16]. In the case of niobium, most works studying the effect of this carbide-forming element are limited to amounts of less than 3% [11,12,17–23] but up to 5% have been also produced along with high amounts of Mo and W [24]. Niobium forms primary carbides in the liquid and are later enclosed by austenite upon solidification; the size and distribution of which depends on the solidification rate (thickness of the casting). Thin castings may produce small well distributed NbC while thick castings will produce large segregated NbC carbides. The wear behavior of these Nb alloyed irons will depend the amount, size and distribution of these carbides which are much harder than the eutectic M_7C_3 (2400 HV and 1500 HV respectively) [25]. The present work analyzes the effect of 2% and 4% Nb in a 15% Cr–3% C iron cast into 45 mm diameter bars, and the abrasive wear behavior is described as a function of the NbC content and the applied load.

2. Experimental Procedure

The experimental alloys for the present work were melted in an open induction furnace by using high purity raw materials. Three 15% Cr–3% C irons with 0%, 2% and 4% Nb were cast at 1500 °C into sand molds to obtain bars of 45 mm diameter. Chemical analysis was undertaken by spectrometry from chill samples obtained during casting for each alloy, and the results are shown in the next section.

The solidification sequence and the final microstructure for each iron was predicted by JMatPro® version 9.0 (Sente Software LTD, Guildford, UK) and verified during the subsequent characterization. For the three alloys, a destabilization heat treatment was undertaken at 900 °C for 30 min followed for an air cooling to room temperature. Samples for metallographic characterization and wear tests, were undertaken from the bars in the as-cast and heat treated conditions. Sample cutting was carried out by mean of abrasive discs; cutting was driven as slow as possible in order to avoid excessive overheating that may cause cracking in the samples. Additionally, copious amounts of water as coolant were used.

Samples for metallographic characterization were prepared in the traditional way of abrasive paper and then polished on nylon cloths by diamond paste (6 μm and then 1 μm). Once polished, the specimens were etched with Villela's reagent (5 mL HCl and 1 g picric acid in 100 mL ethanol) for 30 s to reveal the microstructure. Carbide volume fraction was measured by image analysis on digital micrographs from deep etched samples. Phase constitution was undertaken by X-ray diffraction (XRD) in a BRUKER D8 (Bruker, Karlsruhe, Germany), diffractometer by using Co-kα radiation in a 2θ range of 30°–130°. An SEM JEOL 6400 (JEOL LTD, Peabody, MA, USA) was also used for imaging by secondary and backscattered electrons. The wear behavior of the high- chromium irons in the as-cast and heat treated conditions was undertaken under abrasive wear by using a rubber wheel testing machine and silica sand as abrasive.

The abrasive wear tests were undertaken for the different irons according to the ASTM G65 standard in both as-cast and heat-treated conditions under three loads (58, 75 and 93 N). were done by placing the rubbing surface of the wheel against a 25×25 mm^2 surface of the iron sample and pouring the abrasive particles between the surfaces at a flow of 3.5 cm^3 s^{-1}. Figure 1 shows a schematic draw of the wear test; and Figure 2 shows two SEM micrographs of the abrasive sand used for the tests to evidence size and shape; the mean size of the abrasive particles was 185 μm. The rubbing surfaces were sliding during 15 min at a speed of 500 rpm. The iron samples were polished to a roughness of 0.25 μm before the test. Three tests were undertaken for each condition and the mean value was plotted. The volume loss for each sample was measured by an optical profiler (NANOVEA

PS50 3D Non-Contact Profiler, Irvine, CA, USA). Worn surfaces and worn surface cross-sections were characterized by Scanning Electron Microscopy (SEM).

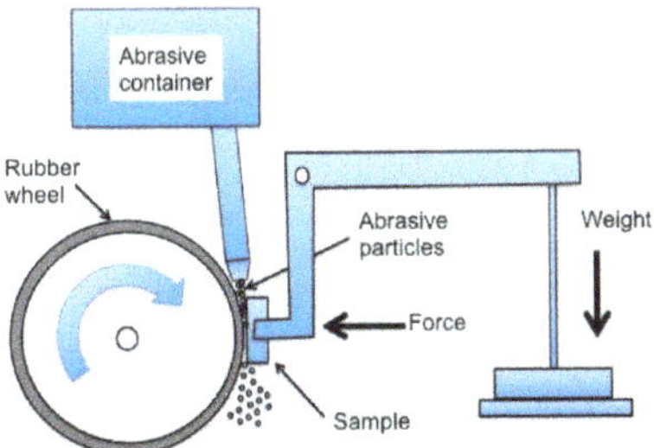

Figure 1. Schematic draw of the abrasive wear test undertaken for this study.

Figure 2. Scanning Electron Microscopy (SEM) micrographs showing the shape and size of the abrasive sand particles used for the wear tests. (**a**) 500 µm, (**b**) 200 µm. Note the presence of sharp particles of a mean size of 185 µm.

3. Results and Discussion

Table 1 shows the chemical composition for the three white cast irons. During making Iron 2 and Iron 3, the niobium additions (as ferro-niobium) were done at the last stage of the melting process before pouring the liquid into the sand molds. Since the ferro-niobium additions were done in a 5 kg based liquid alloy, for higher additions of Fe–Nb less amount of the other elements were expected. This can be seen from Table 1, most element content decreases as the niobium content increased. Niobium contents were 0.014% for Iron 1, 1.795% for Iron 2 and 3.983% for Iron 3; variations in the rest of the elements are considered small for the behavior of the alloys and the effect will focus on the niobium content to describe the alloys behavior.

Table 1. Chemical composition for the three niobium-added irons.

Alloy	%C	%Cr	%Mo	%Ni	%Si	%Mn	%Nb	%Fe
Iron 1	3.12	15.10	3.02	0.56	0.532	1.11	0.014	Balance
Iron 2	3.03	14.89	2.91	0.53	0.492	1.00	1.795	Balance
Iron 3	2.82	14.63	2.88	0.89	0.512	0.93	3.983	Balance

3.1. As-Cast Microstructure

Figure 3 shows the solidification sequence as predicted by JMatPro for two of the experimental irons; (a) Iron 1 (No Nb), and (b) Iron 3 (3.98% Nb). The solidification path is as follows: for Iron 1 solidification starts with the formation of austenite dendrites followed by the eutectic austenite/M_7C_3 and at the end of the solidification process, the formation of small amounts of carbide type M_2C

(Molybdenum carbide). In the case of the Iron 3, the first phase to solidify is MC (niobium carbide) then the sequence is the same than that for the Iron I, formation of austenite dendrites, the eutectic austenite/M_7C_3 and finally small amounts of M_2C. Therefore, the predicted microstructure for the base iron (Iron 1) is composed by a network of carbides M_7C_3 in a matrix of austenite with some M_2C carbides, and for the case of the irons with niobium additions, the microstructure is the same as that for the Iron 1 but the difference is the presence of the NbC phase within the austenitic matrix.

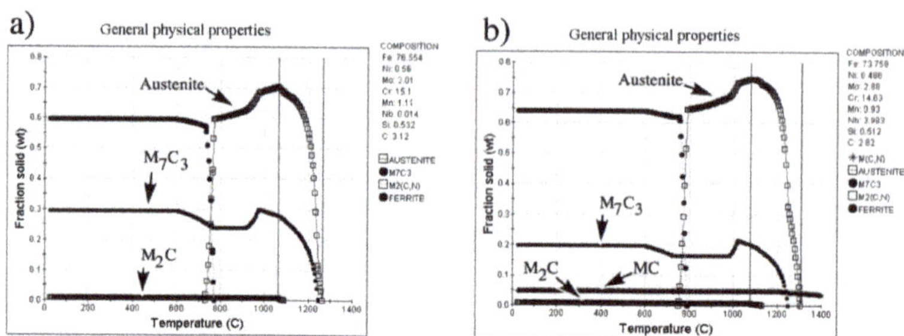

Figure 3. Phases predicted by JMatPro for the irons. (**a**) no-niobium additions (**b**) the iron with 3.98% Nb.

Figure 4 shows the as-cast microstructure for the three experimental irons. Note the presence of NbC in the matrix for the irons with 1.79% and 3.98% Nb, and that the amount and size of these carbides increase with the niobium content. These SEM micrographs show the real microstructure and Table 2 gives the measured amount of each phase and they are compared with the prediction of JMatPro. Note that the predicted and measured amounts of each phase are close, but a small difference was noticed since the real material does not solidify under equilibrium conditions as predicted by the software. The observed microstructure has been widely reported by several authors [12,17–23] for niobium alloyed white irons. For these measurements, the matrix was considered to be fully austenitic; however, it is well known that some martensitic transformation follows the cooling down process after solidification. Such a transformation takes place at the eutectic austenite and/or at the interface pro-eutectic austenite/eutectic carbide. Figure 5 shows evidence of the presence of martensite and M_2C carbide in the 3.98% Nb iron; similar martensitic transformations were also observed for the other as-cast irons. It has been widely reported [1,3,9] that during cooling down to room temperature the austenite close to the eutectic carbide gets impoverished in carbon and it is prone to transform to martensite [1]. The presence of molybdenum rich carbide M_2C has been reported in white irons containing Mo, since this element partitions partially to the matrix, partially to the M_7C_3 carbide and also forms M_2C [26–29].

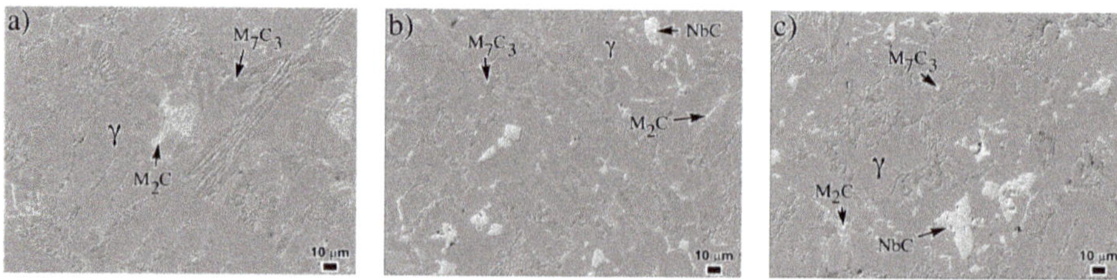

Figure 4. SEM micrographs showing the as-cast microstructure for the three experimental irons; (**a**) 0% Nb, (**b**) 1.79% Nb, and (**c**) 3.98% Nb.

Table 2. Phases measured and predicted by JMatPro® for the three experimental irons.

Alloy		Austenite		M_7C_3		NbC		M_2C	
		Measured	Predicted	Measured	Predicted	Measured	Predicted	Measured	Predicted
Iron 1	(0.01%Nb)	64%	70%	30%	29%	-	-	2%	1%
Iron 2	(1.79%Nb)	65%	72%	26%	24%	3%	2%	1%	1%
Iron 3	(3.98%Nb)	65%	74%	24%	20%	6%	5%	1%	1%

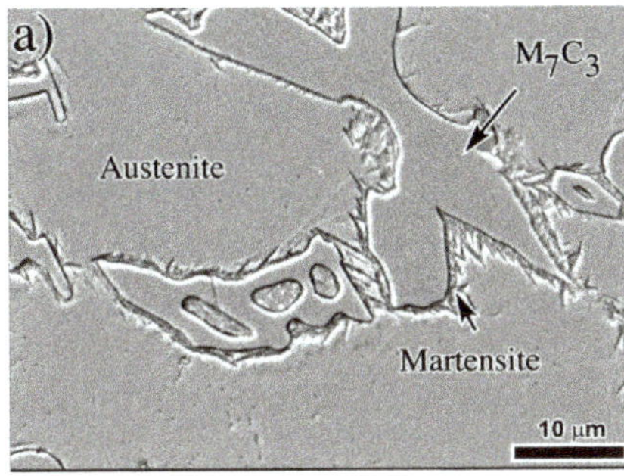

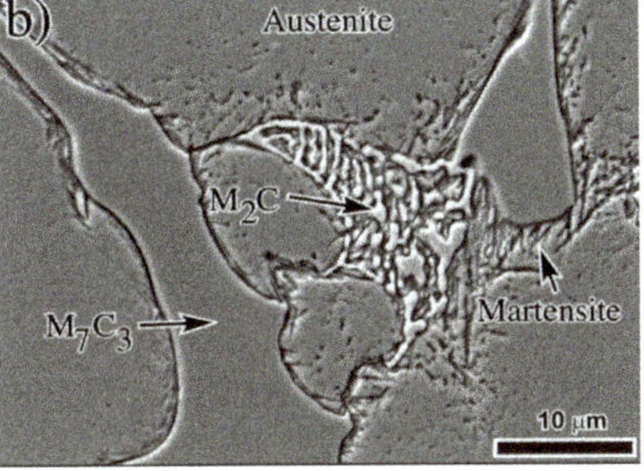

Figure 5. SEM micrographs showing evidence of the presence of martensite (**a**), and M_2C (**b**) in the microstructure of the 3.98% Nb as-cast iron.

3.2. As-Heat Treated Microstructure

The commonly applied heat treatment to destabilize the austenitic matrix in these irons, involves holding at a temperatures usually between 900 and 1000 °C for 1–6 h depending on the size of the casting. During soaking at these temperatures, secondary carbides precipitate in matrix reducing its alloy content, particularly, the carbon content. The reduced alloy content of the austenitic matrix increases the M_S temperature, so that, on cooling to room temperature, the matrix is likely to transform to martensite. Air-cooling from the destabilization temperature is usually sufficient to produce a predominantly martensitic structure while avoiding quench cracking. For the present case, the irons were heat treated at 900 °C for 30 min. The resulted matrix structure was composed of a mixture of retained austenite, martensite and secondary carbides. These microstructures can be seen from SEM pictures from Figure 6 and detail of the nano-sized secondary carbides (SC) can be observed in Figure 7. The amounts of martensite and retained austenite were calculated according to the procedure described by Kim [30] by using X-ray diffraction data and the results are shown from Table 3 where the as-cast phases content are also included, and Figure 8 shows hardness values for both as-cast and heat treated conditions.

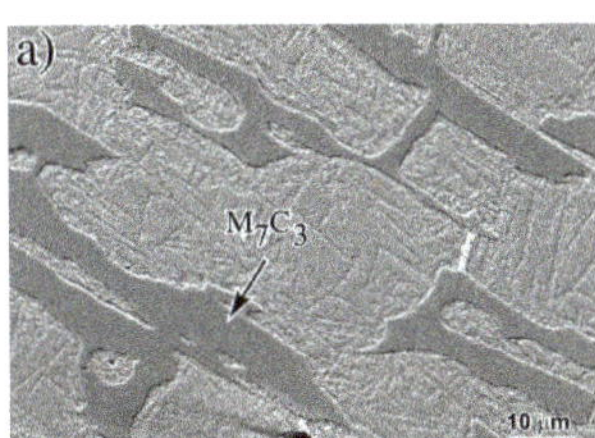

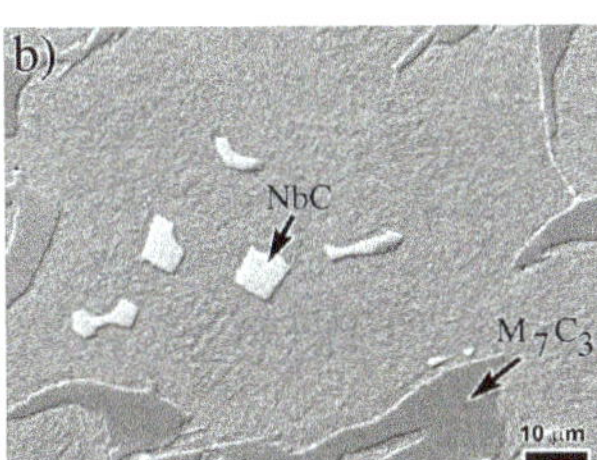

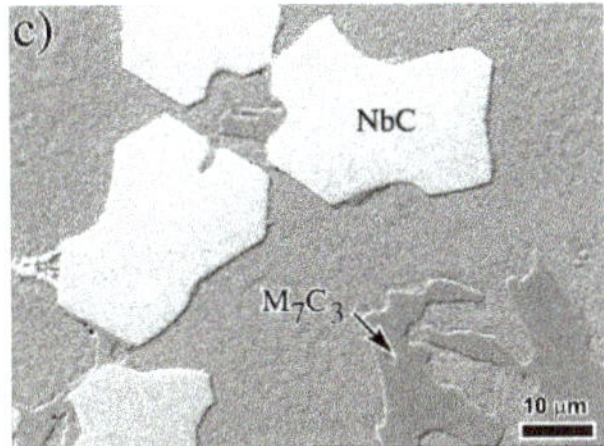

Figure 6. SEM micrographs showing the as-heat treated microstructure; (**a**) 0% Nb, (**b**) 1.79% Nb and (**c**) 3.98% Nb.

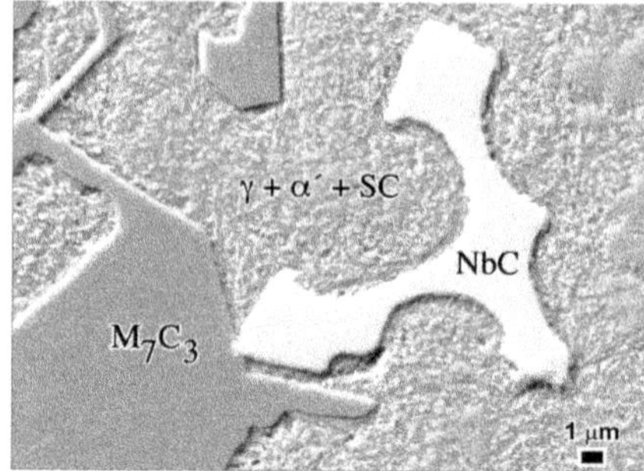

Figure 7. SEM micrograph showing detail of the as-heat treated microstructure of the 4% Nb iron, showing the matrix to be a mixture of retained austenite (γ'), martensite (α') and secondary carbides (SC).

Table 3. Phase content for both as-cast and heat treated conditions for the three irons.

Alloy	Condition	Austenite (%)	Martensite (%)	M_7C_3 (%)	NbC (%)	M_2C (%)
Iron 1	As-Cast	64	4	30	-	2
(0.01%Nb)	Heat Treated	18	50	30	-	2
Iron 2	As-Cast	65	5	26	3	1
(1.79%Nb)	Heat Treated	21	49	26	3	1
Iron 3	As-Cast	65	4	24	6	1
(3.98%Nb)	Heat Treated	23	46	24	6	1

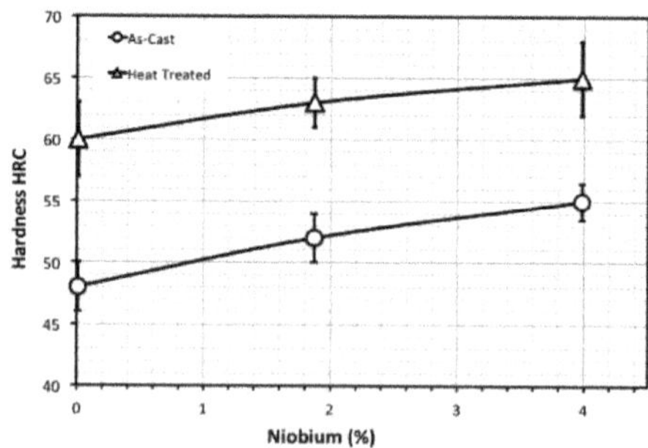

Figure 8. Hardness for the experimental irons as a function of the niobium content for the three irons.

As can be seen, hardness increases as the niobium content increased in the alloys (from 48 to 55 HRC). Furthermore, hardness increased for the heat treated irons and the same tendency was observed when the niobium content increased. The presence of hard NbC contributed to increase hardness of the irons and the precipitation of secondary carbides along with the partial transformation of austenite to martensite in matrix also contributed to increase hardness during the heat treatment (from 60 to 65 HRC). Table 4 shows the microhardness values for the matrix (as-cast and heat treated) and also for the carbides M_7C_3 and NbC. As can be see, the high hardness of the NbC (about 1230 HV) makes a strong contribution to the overall hardness of the iron. Hradness for the M_7C_3 remains unchanged for the different Nb additions (around 1110 HV) but the matrix indeed reduces its hardness hen niobium increased. Such a decrease in hardness is attributed to the smaller amount of carbon in austenite as niobium oncreases; as explained above, niobium consumes high amounts of carbon to form the NbC which promotes a depletion of carbon in matrix.

Table 4. Microhardness HV_{15} of the phases present in the as-cast and heat treated conditions for the three irons.

Alloy	Condition	Matrix	M_7C_3	NbC
Iron 1	As-Cast	291 ± 11	1105 ± 55	-
(0.01%Nb)	Heat Treated	509 ± 19	-	-
Iron 2	As-Cast	271 ± 07	1115 ± 31	2318 ± 31
(1.79%Nb)	Heat Treated	489 ± 14	-	-
Iron 3	As-Cast	258 ± 11	1118 ± 41	2333 ± 36
(3.98%Nb)	Heat Treated	470 ± 05	-	-

3.3. Wear Behavior

Figure 9 shows an example of the topography of the worn surface as generated by the non-contact profiler for the heat treated iron with 3.98% Nb. The volume loss for this particular test was 1.821 mm^3 according to the software used to analyze the worn surface. Each surface after the wear test for each material was analyzed under the profiler, and Figure 10 shows the obtained results of the volume loss for both as-cast and heat treated irons; each point on the plots is the mean value of three tests. As expected, wear increased as the applied load increased for both as-cast and heat treated conditions. In addition, wear decreased as the niobium content increased, which could also be expected since the presence of this element increased the hardness of the overall alloy. However, this behavior was not clear for the tests under the highest applied load (93 N), where the wear behavior seemed to be similar for the different irons in the as-cast conditions; the same behavior was also observed for the heat treated irons. It is suggested that this behavior can be attributed to massive carbide cracking at the very surface and also below the worn surface when the tests were undertaken at the highest load, as explained below.

For the as-cast iron without niobium additions (Figure 10a), the volume lost after the wear test at 58 N is 9.011 mm^3 and it increased to 14.21 mm^3 when the applied load increased to 93 N. For this iron, carbide cracking was observed below the worn surface for all the applied loads. Obviously, the intensity of carbide cracking was more severe for the load of 93N (see Figure 11). For the case of the 3.98%Nb iron, a little of carbide cracking was also observed below the worn surface at 58 N load but only for the M_7C_3 carbide and not for the NbC. On the contrary, at loads of 93 N, severe cracking was observed for both M_7C_3 and NbC carbides (see Figure 12).

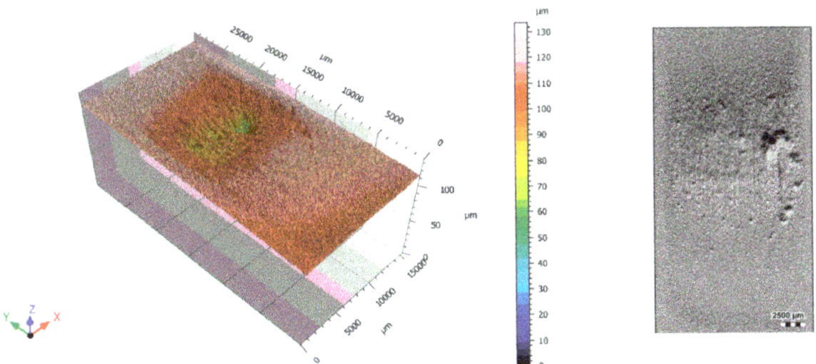

Figure 9. Topography of the worn surface of the 3.98% Nb, heat treated iron, generated by an optical profiler.

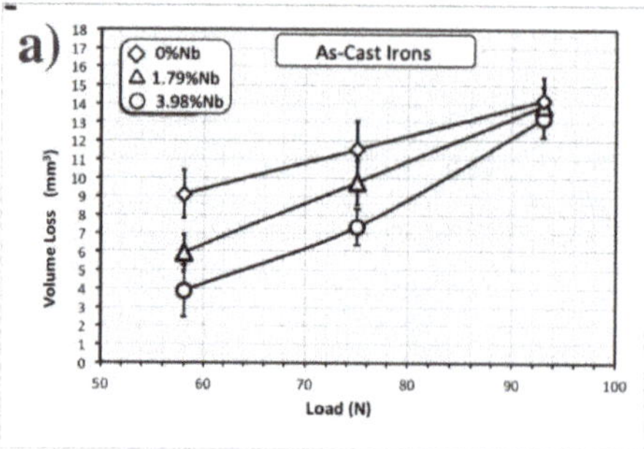
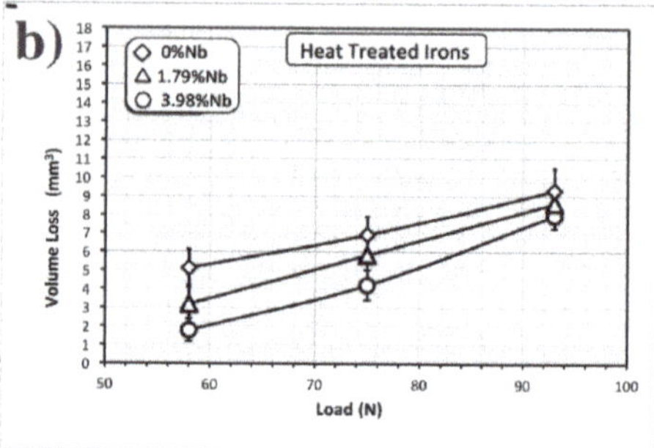

Figure 10. Volume loss load after the wear tests as a function of the applied load for both (**a**) as-cast and (**b**) heat treated irons. Note the increase in volume loss as the load increases, and the lower volume loss for the heat treated irons.

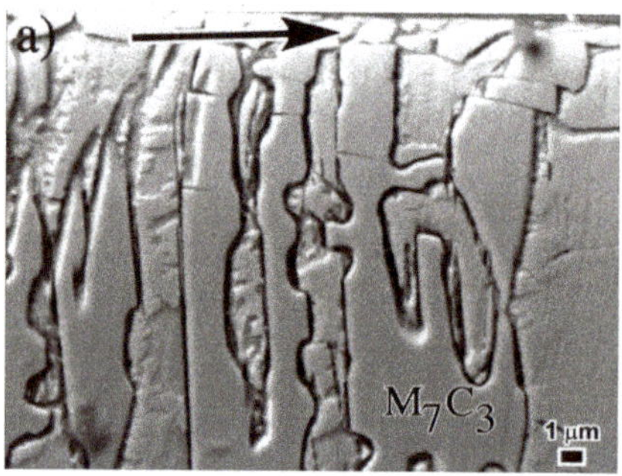
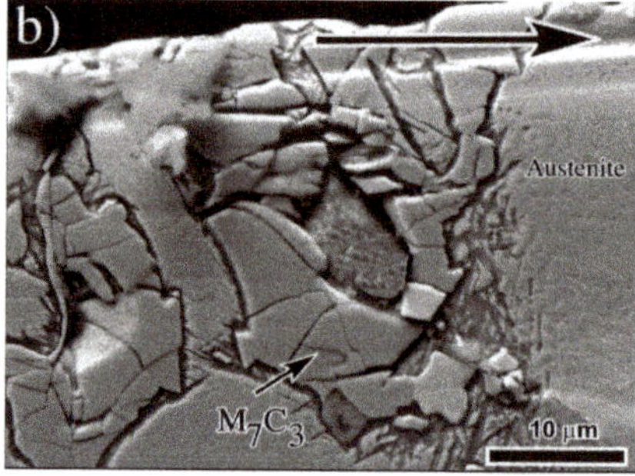

Figure 11. SEM micrographs showing the cross section below the worn surface for the as-cast 0% Nb iron after the wear tests. The arrow on the upper part indicates the sliding direction during abrasion. Note the intensity of carbide cracking for each applied load; (**a**) 58 N, and (**b**) 93 N.

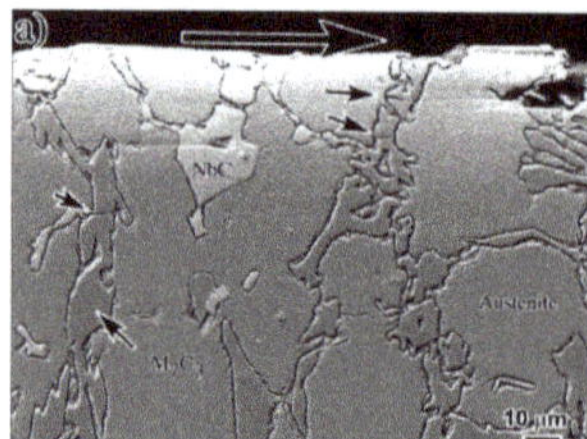
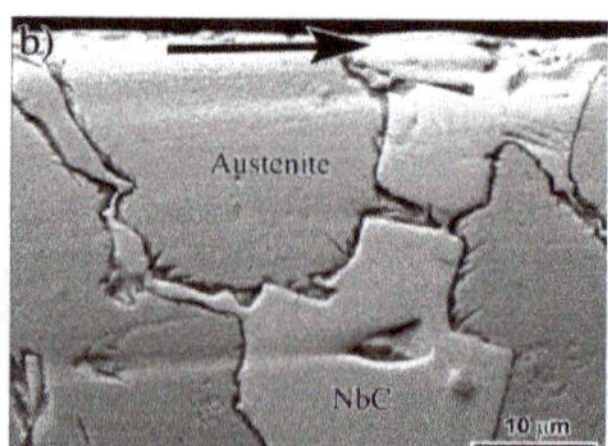
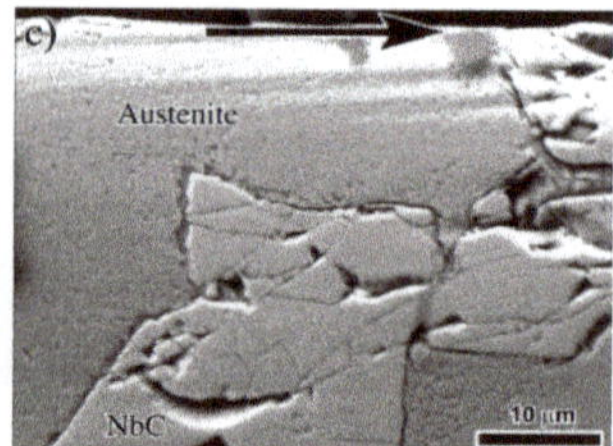

Figure 12. SEM micrographs showing the cross section below the worn surface for the as-cast 3.98% Nb after the wear tests. The arrow on the upper part indicates the sliding direction during abrasion. (**a**) General microstructure showing some cracks at the back of some M_7C_3 carbides (arrowed) for a test load of 58 N. (**b**) Detail of the NbC carbide from Figure a to note there is no cracking on this hard carbide, and (**c**) severe cracking of the NbC for the test load of 93 N.

Carbides cracking at/or below the worn surface has been highlighted by several authors [11,31–33] to be an important cause of surface destabilization during the wear tests. Due to their hardness, carbides are the main responsible for the acceptable wear resistance of these irons; however, if the wear conditions are severe during the test, these carbides may be crushed or fragmented and under these conditions the surface is prone to loss high amounts of material. For the as-cast irons at this high load, the soft austenitic matrix does not offer the adequate support to the carbide phase. High levels of plastic deformation at the surface transfer high stresses to the brittle carbides producing cracking, particularly for large carbides [32]. Figure 13 shows detail of the carbide trituration due to severe plastic deformation of both M_7C_3 and NbC carbides at the surface of the as-cast 3.98% Nb iron.

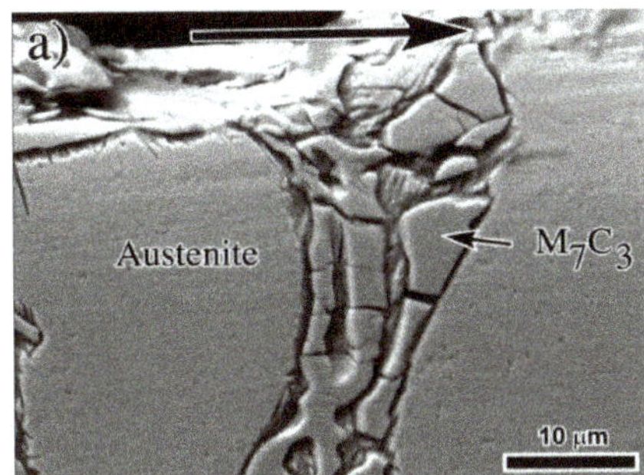

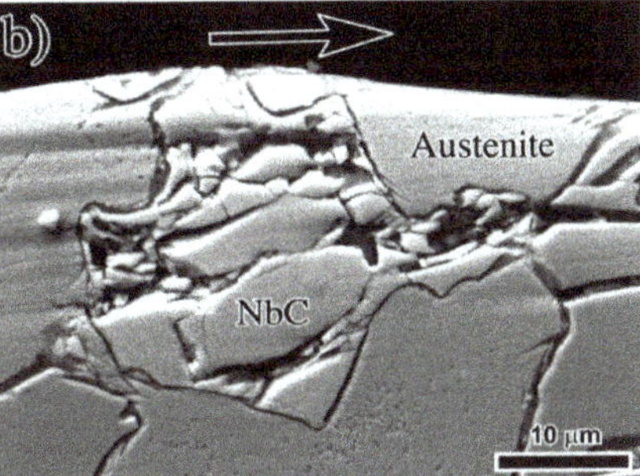

Figure 13. SEM micrographs showing severe carbide cracking below the worn surface for the as-cast 3.98%Nb iron after the wear test at 93 N; (**a**) M_7C_3 carbide, and (**b**) NbC carbide. The arrow indicates the sliding direction.

For the case of the heat treated irons, the volume loss is obviously less than that for the as-cast irons, since the former austenitic matrix has been transformed to a mixture of martensite plus retained austenite plus secondary carbides. Such transformation has caused an increase in hardness of the irons and also an increase in wear resistance as can be seen from Figure 10. Similarly to the as-cast alloys, the volume loss after the wear tests increased with the applied load and decreased with the niobium content. Again, for the highest applied load of 93 N, there was not much difference on the wear behavior for the three irons. For the heat-treated iron without niobium additions (Figure 10b), the volume loss after the wear test at 58 N was 5.015 mm^3 and it increased to 9.521 mm^3 when the applied load increased to 93 N. For these irons, carbide cracking was also observed below the worn surface for all the applied loads, but the intensity of carbide cracking was much less than that observed for the as-cast alloys. For the 3.98%Nb iron, NbC carbide cracking was observed for the samples tested at 75 and 93 N, and not for the iron tested at 58 N. For this later case, just M_7C_3 carbide cracking was detected. Figures 14 and 15 show SEM micrographs of cross sections for the heat-treated iron without niobium (Figure 14) and for the iron with 3.98% Nb also in the heat-treated conditions (Figure 15). Note the intensity of M_7C_3 and NbC carbide cracking at the different applied loads.

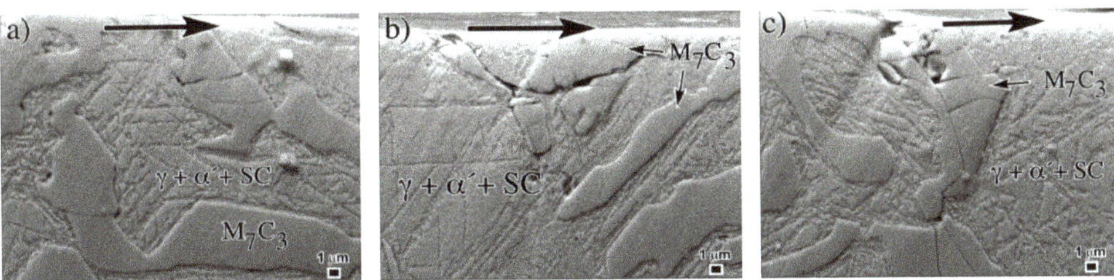

Figure 14. SEM micrographs of cross section of the heat-treated iron without niobium tested at (**a**) 58 N, (**b**) 75 N and (**c**) 93 N. The arrows indicate the sliding direction.

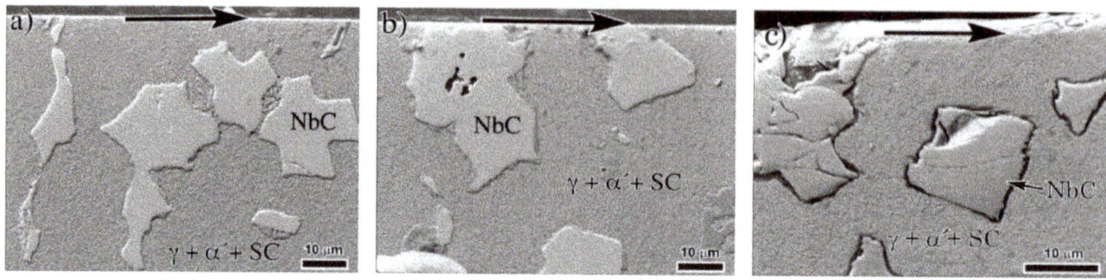

Figure 15. SEM micrographs of cross section of the heat-treated iron with 3.98% niobium tested at (**a**) 58 N, (**b**) 75 N and (**c**) 93 N. The arrows indicate the sliding direction.

Figure 16 shows two SEM micrographs of the worn surface after a deep etching to remove part of the matrix and to evidence the carbide cracking at the very surface in the 3.98% Nb heat-treated iron tested at 93 N.

Figure 16. SEM micrographs of the worn surface showing carbide cracking; (**a**) general view, and (**b**) detail of cracks on a NbC carbide. Heat treated iron with 3.98% Nb tested at 93 N. The large arrows indicate the sliding direction.

From these observations, it is evident again that carbide cracking at/or below the worn surface determines the wear behavior of the irons. Previous works on silicon-alloyed irons [34] and titanium-alloyed irons [33] have reported a linear relationship between wear rate and deep of deformation under pure sliding wear tests. Such a deep of deformation is also related to the depth at which carbide cracking occurs below the worn surface. Therefore, as highlighted by Fulcher et al. [32], the role of the matrix should be to protect the carbides against bending due to the absorbed stresses during the wear tests. For the heat treated irons, the matrix composed by martensite plus retained austenite plus secondary carbides provides better support against carbide cracking than that offered by the austenitic matrix in the as-cast irons.

Figure 17 shows plots of the wear intensity or specific wear rate for the experimental irons in both as-cast (Figure 17a) and heat-treated (Figure 17b) conditions. As can be seen, the wear rate is a constant for any load for the iron without niobium. At any load, the wear rate is the same since carbide cracking was observed for all the tests on these irons. However, for the irons with niobium additions, the wear rate increased with load and decreased with the niobium content. This behavior also evidence that for high loads the presence of hard NbC carbides does not offer too much wear resistance due to carbide cracking at these high loads.

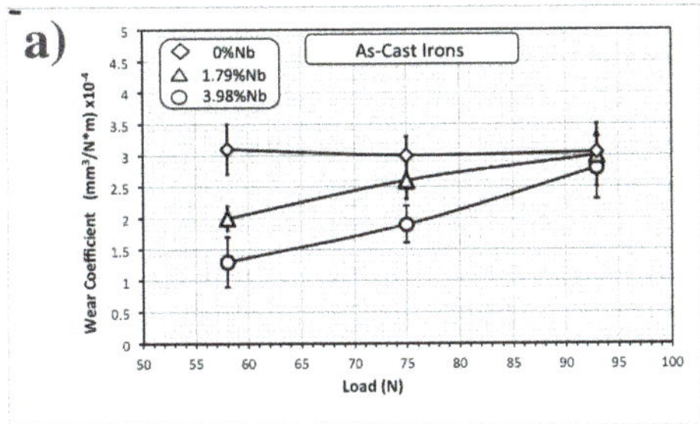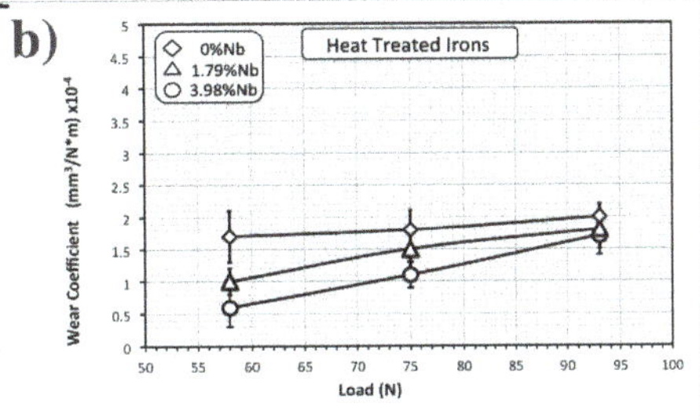

Figure 17. Wear coefficient against load for the (**a**) as-cast and (**b**) heat-treated irons. Note the lower wear coefficient for the heat treated irons and that the niobium content makes no difference for the highest loads used in this work.

4. Conclusions

For the as-cast conditions, niobium additions increased the hard NbC phase since since it is a strong carbide-forming element. This caused a decrease in the eutectic M_7C_3 volume fraction from 30% for the iron without niobium to 24% for the iron with 3.98% Nb. However, NbC increased to 6%. The overall carbide volume fraction in the irons was about 30% but the presence of NbC in the Nb-added irons increased hardness from 48 to 55 HRC.

After a destabilization heat treatment at 900 °C, secondary carbides precipitation occurred within the matrix. The matrix then partially transformed to martensite during the subsequent cooling down. The presence of martensite and secondary carbides in the matrix increased the overall hardness of the irons from 60 to 65 HRC as the Nb content increased.

The abrasive wear resistance increased with Nb for both as-cast and heat-treated alloys due to the increase in hardness. However, not much difference was noticed for the high load of 93 N, which is attributed to the severe carbide cracking at and below the worn surface. Wear resistance also increased for the heat-treated alloys due to the matrix strengthening by martensite and secondary carbides. Wear intensity was constant at any load for the iron without niobium, while it increased with the Nb content for both as-cast and heat-treated conditions.

Author Contributions: A.B.-J. and F.G. wrote the paper and conceived the experiments. U.V. performed the experiments. I.M. analyzed the data software.

Funding: Grant: 258873, National Council of Science and Technology of Mexico (CONACyT).

Aknowledgments: This paper results from the financial support of the CONACyT from Mexico under the grant No: 258873 on the announcement of Basic Science 2015. The authors are very grateful to the Universidad Michoacana of Mexico for the complimentary support. One of the authors, Uzzi Vera, also acknowledges CONACyT for the scholarship during his undergrade studies.

Conflicts of Interest: The authors declare no conflict of interest.

References

1. Tabrett, C.P.; Sare, I.R.; Gomashchi, M.R. Microstructure-property relationships in high-chromium white iron alloys. *Int. Mater. Rev.* **1996**, *41*, 59–82. [CrossRef]
2. Albertin, E.; Sinatora, A. Effect of carbide fraction and matrix microstructure on the wear of cast iron balls tested in a laboratory ball mill. *Wear* **2001**, *250*, 492–501. [CrossRef]
3. Kositsyana, I.I.; Sagaradze, V.V.; Makarov, A.V.; Kozlova, A.N.; Ustyuzhaninova, A.I. Effect of structure on the properties of white chromium cast irons. *Mater. Sci. Heat Treat.* **1996**, *38*, 149–152. [CrossRef]
4. Filippov, M.A.; Lhagvadorzh, P.; Plotnikov, G.N. Structural factors of elevation of wear-resistance of white chromium cast iron. *Mater. Sci. Heat Treat.* **2000**, *42*, 423–426. [CrossRef]
5. Wang, J.; Li, C.; Liu, H.; Yang, H.; Shen, B.; Gao, S.; Huang, S. The precipitation and transformation of secondary carbides in a high chromium cast iron. *Mater. Charact.* **2006**, *56*, 73–78. [CrossRef]

6. Kootsokoos, A.; Gates, J.D. The role of secondary carbide precipitation on the fracture toughness of a reduced carbon white iron. *Mater. Sci. Eng. A* **2008**, *480*, 313–318. [CrossRef]

7. Powell, G.L.; Bee, J.V. Secondary carbide precipitation in a 18 wt%Cr-1 wt% Mo white iron. *J. Mater. Sci.* **1996**, *31*, 707–711. [CrossRef]

8. Wang, J.; Sun, Z.; Zuo, R.; Li, C.; Shen, B.; Gao, S.; Huang, S. Effects of secondary carbide precipitation and transformation on abrasion resistance of the 16Cr-1Mo-1Cu white iron. *J. Mater. Eng. Perform.* **2006**, *15*, 316–319. [CrossRef]

9. Bedolla-Jacuinde, A.; Arias, L.; Hernández, B. Kinetics of secondary carbides precipitation in a high-chromium white iron. *J. Mater. Eng. Perform.* **2003**, *12*, 371–382. [CrossRef]

10. Bedolla-Jacuinde, A.; Guerra, F.V.; Mejía, I.; Zuno-Silva, J.; Maldonado, C. Boron effect of precipitation of secondary carbides during destabilization of a high-chromium white iron. *Int. J. Cast Met. Res.* **2016**, *29*, 55–61. [CrossRef]

11. Bedolla-Jacuinde, A.; Guerra, F.V.; Mejía, I.; Zuno-Silva, J.; Rainforth, M.W. Abrasive wear of V-Nb-Ti alloyed high-chormium white irons. *Wear* **2015**, *332*, 1006–1011. [CrossRef]

12. Bedolla-Jacuinde, A. Microstructure of vanadium-, niobium-, and titanium-alloyed high-chromium white irons. *Int. J. Cast Met. Res.* **2001**, *13*, 343–361. [CrossRef]

13. Bedolla-Jacuinde, A.; Correa, R.; Quezada, J.G.; Maldonado, C. Effect of titanium on the as-cast microstructure of a 16% chromium white iron. *Mater. Sci. Eng. A* **2005**, *297*, 297–308. [CrossRef]

14. Fras, E.; Kawalec, M.; López, H.F. Solidification microstructures and mechanical properties of high-vanadium Fe-C-V and Fe-C-V Si alloys. *Mater. Sci. Eng. A* **2009**, *254*, 193–203. [CrossRef]

15. Filipovic, M.; Kamberovic, Z.; Korak, M. Solidification of high chromium white cast iron alloyed with vanadium. *Mater. Trans.* **2011**, *52*, 386–390. [CrossRef]

16. Efremenko, V.G.; Shimizu, K.; Cheiliakh, A.P.; Kozarevskaya, T.V.; Kusumoto, K.; Yamamoto, K. Effect of vanadium and chromium on the microstructural features of V-Cr-Mn-Ni spheroidal carbide cast irons. *Int. J. Miner. Metall. Mater.* **2014**, *21*, 1096–1108. [CrossRef]

17. Zhi, X.; Xiang, J.; Fu, H.; Xiao, B. Effect of niobium on the as-cast microstructure of hypereutectic high chromium cast iron. *Mater. Lett.* **2008**, *62*, 857–860. [CrossRef]

18. Penagos, J.J.; Pereira, J.J.; Machado, P.C.; Albertin, E.; Sinatora, A. Synergetic effect of niobium and molybdenum on abrasion resistance of high chromium cast irons. *Wear* **2017**, *376*, 983–992. [CrossRef]

19. Filipovic, M.; Kemberovic, Z.; Korac, M.; Gavrilovski, M. Microstructure and mechanical properties of Fe-Cr-C-Nb white cast irons. *Mater. Des.* **2013**, *47*, 41–48. [CrossRef]

20. Filipovic, M.; Kamberovic, M.; Korac, M.; Jordovic, B. Effect of niobium and vanadium additions on the as-cast microstructure and properties of hypoeutectic Fe-Cr-C alloy. *ISIJ Int.* **2013**, *53*, 2160–2166. [CrossRef]

21. Chen, H.X.; Chang, Z.C.; Lu, J.C.; Lin, H.T. Effect of niobium on wear resistance of 15%Cr white cast iron. *Wear* **1993**, *166*, 197–201.

22. Bouhamla, K.; Hadji, A.; Maouche, H.; Merradi, H. Effect of niobium on wear resistance of heat treated high chromium cast iron. *Revue de Métallurgie* **2011**, *108*, 83–88. [CrossRef]

23. Fiset, M.; Peev, K.; Radulovic, M. The influence of niobium on fracture toughness and abrasion resistance in high chromium white cast irons. *J. Mater. Sci. Lett.* **1993**, *12*, 615–617. [CrossRef]

24. Kusumoto, K.; Shimizu, K.; Yaer, X.; Zhang, Y.; Ota, Y.; Ito, J. Abrasive wear characteristics of Fe-2C-5Cr-5Mo-5W-5Nb multi-component white cast iron. *Wear* **2017**, *376*, 22–29. [CrossRef]

25. Dogan, O.; Hawk, A.; Tilkzac, H. Wear of cast chromium steels with TiC reinforcement. *Wear* **2001**, *250*, 462–469. [CrossRef]

26. Loper, C.R.; Baik, H.K. Influence of molybdenum and titanium on the microstructures of Fe-C-Cr-Nb white cast irons. *AFS Trans.* **1989**, *97*, 1001–1008.

27. Maratray, F. Choice of appropriate compositions for chromium-molybdenum white irons. *AFS Trans.* **1971**, *79*, 121–124.

28. DeMello, J.D.B.; Duran-Charre, M.; Hamar-Thibualt, S. Solidification and solid state transformations during cooling of chromium-molybdenum white cast irons. *Metall. Trans. A* **1983**, *14*, 1793–1801. [CrossRef]

29. Ikeda, M.; Umeda, T.; Tong, C.P.; Suzuki, T.; Niwa, N.; Kato, O. Effect of molybdenum addition on solidification structure, mechanical properties and wear resistivity of high chromium cast iron. *ISIJ Int.* **1992**, *32*, 1157–1162. [CrossRef]

30. Kim, C. X-ray method of measuring retained austenite in heat treated white cast irons. *J. Heat Treat.* **1979**, *1*, 43–51. [CrossRef]

31. Radulovic, M.; Fiset, M.; Peev, K.; Tomovic, M. The influence of vanadium on fracture and abrasion resistance in high chromium white cast irons. *J. Mater. Sci.* **1994**, *29*, 5085–5094. [CrossRef]

32. Fulcher, J.K.; Kosel, T.H.; Fiore, N.F. The effect of carbide volume fraction on the low stress abrasion resistance of high Cr-Mo white cast irons. *Wear* **1983**, *84*, 313–325. [CrossRef]

33. Bedolla-Jacuinde, A.; Correa, R.; Mejía, I.; Quezada, J.; Rainforth, W.M. The effect of titanium on the wear behavior of a 16%Cr white cast iron under pure sliding. *Wear* **2007**, *263*, 808–820. [CrossRef]

34. Bedolla-Jacuinde, A.; Rainforth, W.M. The wear behavior of high-chromium white cast irons as a function of silicon and Mischmetal content. *Wear* **2001**, *250*, 449–461. [CrossRef]

Article

A Comparative Study on the Influence of Chromium on the Phase Fraction and Elemental Distribution in As-Cast High Chromium Cast Irons: Simulation vs. Experimentation

U. Pranav Nayak [1,*], María Agustina Guitar [1] and Frank Mücklich [1,2]

1   Department of Materials Science, Saarland University. Campus D3.3, D-66123 Saarbrücken, Germany; a.guitar@mx.uni-saarland.de (M.A.G.); muecke@matsci.uni-sb.de (F.M.)
2   Materials Engineering Center Saarland (MECS), Campus D3.3, D-66123 Saarbrücken, Germany
*   Correspondence: pranav.nayak@uni-saarland.de; Tel.: +49-681-302-70517

Received: 21 November 2019; Accepted: 19 December 2019; Published: 23 December 2019

Abstract: The excellent abrasion resistance of high chromium cast irons (HCCIs) stems from the dispersion of the hard iron-chromium eutectic carbides. The surrounding matrix on the other hand, provides sufficient mechanical support, improving the resistance to cracking deformation and spalling. Prior knowledge of the microstructural characteristics is imperative to appropriately design subsequent heat treatments, and in this regard, employing computational tools is the current trend. In this work, computational and experimental results were correlated with the aim of validating the usage of MatCalc simulations to predict the eutectic carbide phase fraction and the elemental distribution in two HCCI alloys, in the as-cast condition. Microstructural observations were carried out using optical microscopy and SEM. The chemical composition and fraction of each phase was measured by electron probe microanalysis and image analysis, respectively. In all cases, the values predicted by the pseudo-equilibrium diagrams, computed with MatCalc, were in accordance with the experimentally determined values. Consequently, the results suggest that time and resource intensive experimental procedures can be replaced by simulation techniques to determine the phase fraction and especially, the individual phase compositions in the as-cast state.

Keywords: high chromium cast irons; eutectic carbide; carbide volume fraction; chemical composition; image analysis; simulation; MatCalc; hardness

---

1. Introduction

High chromium cast irons (HCCIs) are alloys containing 15–30 wt. % Cr and 2.5–4 wt. % C, and belong to the Fe–C–Cr ternary system, as described in the ASTM A532 [1,2]. Other international standards such as ISO 21988:2006(E) classifies HCCIs under five different grades with Cr contents ranging from 11 to 40 wt. % [3]. They primarily contain hard eutectic carbides (EC) of the M_7C_3 type dispersed in a supportive, modifiable matrix. Although equilibrium solidification would result in a ferritic matrix, the final microstructure primarily contains an austenitic matrix indicating a non-equilibrium nature of solidification. The M_7C_3 carbides contribute to the hardness and wear resistance whereas the relatively softer matrix helps in improving the toughness of the HCCI alloy. This combination makes it an attractive choice for usage in applications demanding excellent abrasion and moderate impact resistance, such as ore crushers, pulverizing equipment, ball mill liners etc., in coal and mineral industries [4,5].

Although the matrix of an HCCI alloy can be modified by employing proper heat treatment, the EC are relatively immune to it [4,6]. Once the EC are formed during solidification, the only way to modify them is by re-melting the cast and chemically modifying the melt by alloying and/or varying the process parameters. Over the years, the influence of several alloying elements such as Ti [7,8], Mo [9,10], W [11,12] etc., on the solidification behavior of the as-cast melt have been assessed. In all cases, the addition of alloying elements modified the eutectic carbide structure which resulted in a change in the final microstructural and mechanical behavior. Initially, the wear resistance was thought to be mainly influenced by the hardness of the material but it is now understood that a lot of factors contribute to this, such as type, volume fraction, size, and morphology of eutectic carbides, and its interaction with the host matrix [13–15]. Not only the carbide characteristics but also the chemical composition of the matrix after solidification determines, to a large extent, the efficacy of the subsequent heat treatment on the microstructural modifications [5,16,17]. For these reasons, it is imperative to evaluate the carbide characteristics and the chemical composition of each phase in the as-cast state.

In order to understand the processes that govern the materials' properties, it is essential to comprehend the phase diagram and phase equilibria for the given alloy composition. Complete phase diagrams for several binary and a few ternary systems are available but their construction becomes cumbersome with every additional element [18]. In this regard, employing computational techniques would be useful to extract valuable thermodynamic properties of the phases rather than experimentation. MatCalc (Materials Calculator), is a thermo-kinetic software package developed for the simulation of precipitation kinetics that occur during various metallurgical processes [19]. It employs the CALPHAD type database, which, currently is the only theoretical approach to carry out thermodynamic and kinetic calculations in multicomponent systems [20].

Studies combining simulation and experimentation have been previously conducted. Li et al. [21] computed the phase diagram of an HCCI alloy containing 15% Cr using Thermo-Calc software and compared the predicted precipitation sequence with differential scanning calorimetry (DSC) measurements. The results were in agreement with each other. Albertin et al. [22] successfully employed computational thermodynamics (Thermo-Calc) in analyzing several different HCCI compositions with the intention of optimizing the hardness and wear resistance after thermal treatments. Akyildiz et al. [23] used MatCalc to simulate pseudo-binary phase diagrams for two HCCI alloys with varying Mo contents. The predicted transformation temperatures were later compared to the DSC values to matching success. Moreover, the increase in the amount of $M_{23}C_6$ carbides with increasing Mo as predicted by the software was also seen in the alloy microstructure.

Having an idea of the microstructural characteristics (carbide volume fraction, phase chemical composition etc.) beforehand would be strongly beneficial in the development of the alloy and an appropriate design of the subsequent heat treatment (HT) to maximize its potential. Therefore, a thorough characterization in the as-cast state will serve as a terminus a quo for further microstructural modifications combining thermodynamic and kinetic calculations, and experimentally performing the thermal treatments.

The main objective of the current work is to validate the usage of thermodynamic simulation as an approach to determine the eutectic carbide phase fraction and the corresponding matrix, and carbide chemical compositions of HCCIs in the as-cast condition. Accordingly, two HCCI compositions (16% Cr and 26% Cr) were manufactured under similar conditions and their corresponding microstructural characterization was carried out in the as-cast state. The bulk compositions of the two materials were used as the input in the simulation software, MatCalc 6, to estimate the phase fraction and the chemical composition of the matrix and carbides, post solidification. Moreover, predictions from Matcalc were experimentally validated by image analysis (I-A) and electron probe microanalysis (EPMA). Additionally, the influence of the Cr content in the hardness of each phase and of the 'composite' itself, was evaluated using nanoindentation, Rockwell and Vickers microhardness tests.

2. Materials and Methodology

High chromium cast iron samples with varying Cr contents (approximately 16% and 26%) were manufactured in an arc furnace and casted at ~1450 °C into rectangular (Y) shaped sand molds hardened with phenolic resin. Test samples were cut from the lower half of the test block measuring

$175 \times 90 \times 25$ mm^3 to ensure they were free from casting defects. Optical emission spectroscopy (GNR Metal Lab 75/80; G.N.R. S.r.l., Agrate Conturbia, Novara, Italy) was used to determine the bulk chemical composition of the castings. Table 1 represents the chemical composition (wt. %) along with the Cr/C ratio of both alloys. The cast samples were cut into pieces measuring $20 \times 20 \times 10$ mm^3 using an abrasive disc and later embedded for microstructural characterization. Standard metallographic procedure was followed as described in [24].

Table 1. Bulk chemical composition (in wt. %) of the samples measured by optical emission spectroscopy. HCCI: high chromium cast irons.

Alloy	C	Cr	Mn	Ni	Mo	Si	Cu	P	S	Fe	Cr/C
16% HCCI (Sample A)	2.43	15.84	0.76	0.18	0.41	0.47	0.04	0.02	0.02	Bal.	6.5
26% HCCI (Sample B)	2.53	26.60	0.66	0.26	0.24	0.37	0.03	<0.01	0.04	Bal.	10.5

In addition to the bulk composition, the individual matrix and carbide compositions was measured by EPMA (8900 R JEOL Superprobe; JEOL, Tokyo, Japan) and the mean of 10 readings was considered. Phase identification was performed using a PANalytical Empyrean X-ray diffractometer (PANalytical B.V., Almelo, The Netherlands) coupled with a Co source, an acceleration voltage of 40 kV and a 40 mA tube current.

The polished samples were etched with three different etchants depending on the objective. Villella's reagent (1 g picric acid + 5 mL HCl + 95 mL C$_2$H$_5$OH) for general microstructure revelation [25], a modified Murakami's reagent (4 g K$_3$[Fe(CN)$_6$] + 8 g KOH + 100 mL distilled water) for eutectic carbide volume fraction (% CVF) determination [26] and a solution of 10% HCl in methanol (CH$_3$OH) for deep etching to reveal the three dimensional (3D) structure of the eutectic carbides [27]. The specifics of the etchings is mentioned in Table 2. In all cases, the samples were immersed in the etchant for the appropriate time, rinsed with water and ethanol, and air dried.

Table 2. Etching parameters for each etchant.

Etchant	Solution Temperature	Etching Time
Vilella's	Room temperature (20 °C)	7–15 s
Modified Murakami's	60 °C	5 min
10% HCl in methanol	Room temperature (20 °C)	24 h

Microscopy observations were carried out using a LEXT OLS 4100 Olympus confocal laser scanning microscope (CLSM) (Olympus Corporation, Tokyo, Japan). It uses a laser with 405 nm wavelength and, a lateral and vertical resolution of 120 and 10 nm, respectively. The microstructures of the samples along with the 3D structure of the carbides were analysed using a FEI Helios Nanolab field emission scanning electron microscope (FE-SEM) working with an acceleration voltage of 5–15 kV and a beam current of 1.4 nA. A high sensitivity backscattered electron detector (vCD) was also used in order to obtain a better contrast between the phases. Furthermore, a Leica CTR6000 microscope (Leica Camera AG, Wetzlar, Germany) coupled with a Jenoptik CCD camera (Jenoptik AG, Jena, Germany) was used for image acquisition of the samples for % CVF determination. The % CVF was calculated after a post processing of the images using the image analysis (I-A) software, ImageJ (version 1.52p) (LOCI, UW-Madison, WI, USA) [28]. The analyzed area was the same for all the images and an average of 10–12 micrographs were considered for each sample.

Equilibrium phase diagrams for both alloys along with thermodynamic, and equilibrium phase fraction calculations were computed using MatCalc 6 software (version 6.01) (MatCalc Engineering GmbH, Vienna, Austria) with the thermodynamic database "ME_Fe 1.2". The results were then correlated with the experimentally and numerically determined values obtained from I-A and empirical equations described in [29,30], respectively. Moreover, the chemical composition of each phase was estimated at a certain temperature range and compared with the EPMA results.

The bulk hardness of the samples was measured using the Rockwell hardness method, with a diamond indenter and a load of 150 kgf (HRC) whereas the matrix hardness was determined using the Vickers method. A Struers Dura Scan 50 microhardness tester (Struers Inc., Cleveland, OH, USA) with a load of 0.9807N (HV0.1) was used for this purpose. In both testing methods, the dwell time was 15 s and an average of 15–20 readings was considered. Nano-indentation (Hysitron TI 900 TriboIndenter; Bruker Corporation, Billerica, MA, USA) was used to calculate the hardness of the eutectic carbides (in GPa). A Berkovich tip, with a tip depth of 200 nm was used in displacement mode. A loading/unloading rate of 50 was maintained and the scan size was approximately 30 microns. Each indentation time was 2 min and the values were averaged over 10–12 readings.

3. Results and Discussion

3.1. Phase and Microstructural Analysis

Figure 1 represents the microstructure of the Sample A (16% HCCI) and Sample B (26% HCCI) as observed under CLSM and SEM. Their microstructure consisted of a network of eutectic carbides (EC) dispersed in a matrix of austenite dendrites. The EC was identified to be M_7C_3 which is consistent with previous studies [1,4,31,32].

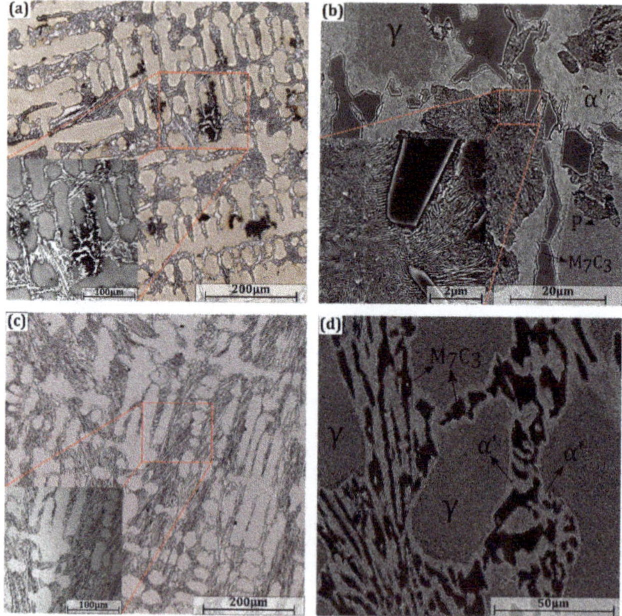

Figure 1. Representative OM (200×) and SEM micrographs of samples A (**a,b**) and B (**c,d**), respectively, after Vilella's etching. Inset in (**a,c**) represents a magnified image (500×) acquired using laser light. The different phases, austenite (γ), martensite (α'), pearlite (P), and eutectic carbides (M_7C_3) are indicated in images (**b,d**). The pearlite phase can be observed in the inset in (**b**).

From the OM and SEM micrographs, it was observed that the matrix of Sample B is completely austenitic whereas some partial transformation to pearlite has occurred in Sample A. Generally, austenitic matrix structures are favored by high Ni and Mo contents, a high Cr/C ratio, and faster cooling rates during casting [33–35]. According to the graph presented by Maratray et al. [33], where the Cr/C ratio and the bulk Mo composition are related to the decomposition of the austenite upon cooling, for a cast alloy with a Mo content of 0.4 wt. %, the Cr/C ratio needs to be around 6.5 (which is the same as Sample A) to avoid austenite decomposition. Therefore, the partial austenite to pearlite transformation in Sample A can be primarily attributed to the low Cr/C ratio (6.5) since both samples, A and B contain negligible amounts of Ni and Mo, were casted under similar conditions, and Sample B presents a Cr/C ratio of 10.5.

In addition to the major phases, austenite and M_7C_3 EC, a thin layer of martensite was observed at the periphery of the carbides, which is clearly visible in the SEM micrographs (Figure 1b,d). The presence of these phases is further confirmed by XRD in Figure 2. The martensite formation is associated with the local C and Cr depletion which takes place during the solidification of the EC in contact with the pro-eutectic austenite, as also observed in other studies [14,24,29]. The impoverishment of alloying elements at the contact zone results in an increase in the martensitic start (M_s) temperature leading to its formation during subsequent cool down.

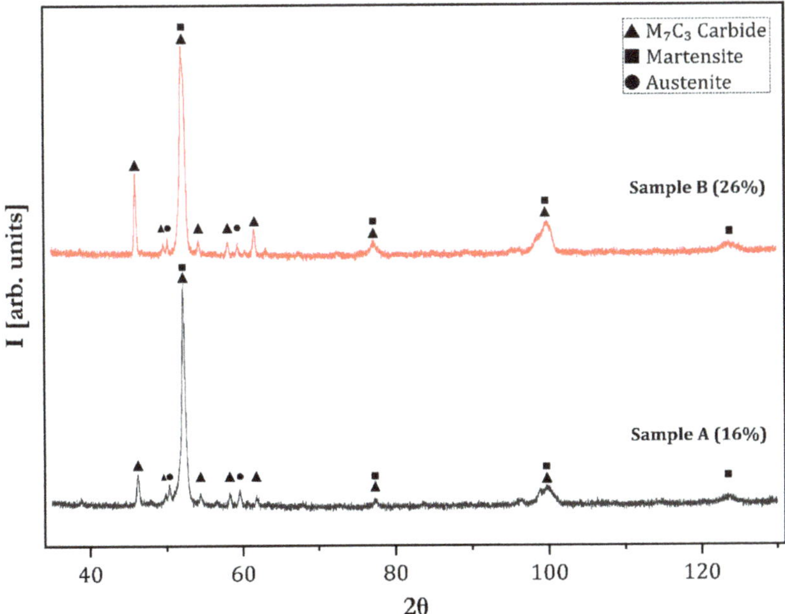

Figure 2. X-ray diffractograms of samples A and B. Austenite, martensite, and the M_7C_3 carbide are indexed for reference.

SEM micrographs depicting the three dimensional structure of the EC are shown in Figure 3. The carbides are located heterogeneously throughout the material possessing a rhombohedral/hexagonal cross section. Their growth mechanism during solidification has been explained elsewhere [31,36,37]. Both rod and plate-like EC are seen as this is a hypoeutectic alloy [15,38]. Moreover, the 'rosette' pattern can be observed in Figure 3a with hollow, fine rods at the center and larger blades as we move away. The difference in their sizes is associated with the decrease in undercooling as the solidification progresses, and the segregation of alloying elements in the melt [39].

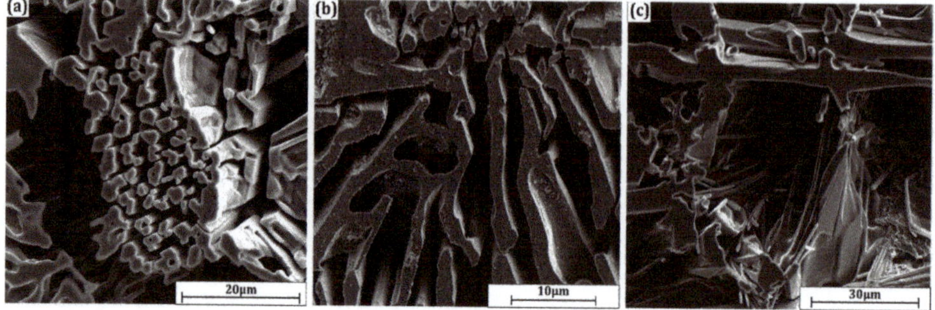

Figure 3. Representative SEM micrographs of Sample B after deep etching. The three-dimensional structure of the eutectic carbide is visible. (**a**) 'rosette' pattern, with the thin hexagonal rods at the center; (**b**) blade morphology; and (**c**) heterogenity in nucleation.

3.2. Pseudo-Binary Phase Diagram

The pseudo-binary phase diagrams for both alloys at carbon contents ranging from 2 wt. % to 5 wt. % were computed using MatCalc as shown in Figure 4. The generation of a pseudo-binary phase diagram will help in understanding the solidification sequence of the alloy and the corresponding equilibrium phase stabilities. This information is essential for alloy development and subsequent heat treatment design. From Figure 4, it can be seen that increasing the bulk chromium content results in an increase in the eutectic transition temperature (1285 → 1315 °C) and a decrease in the eutectic carbon content (3.88 wt. % → 3.24 wt. % C). Despite the similar carbon contents in the alloys, Sample B shows a smaller solidification range (30 °C) compared to its counterpart, Sample A (70 °C). It is due to the addition of chromium which increases the fraction of EC formed and improves its stability [21,23]. As a result, the formation temperature of the EC in Sample B is higher (1302 °C) in comparison with Sample A (1281 °C), indicating that it is stable for a larger range of temperatures compared to the latter. The phase diagram also predicts the formation of an additional carbide, $M_{23}C_6$, in Sample B around 1000 °C which suggests that increasing the Cr/C ratios stabilize the $M_{23}C_6$ carbide. Nevertheless, $M_{23}C_6$ was not experimentally observed, possibly due to faster cooling rate and very low Mo content [10,40].

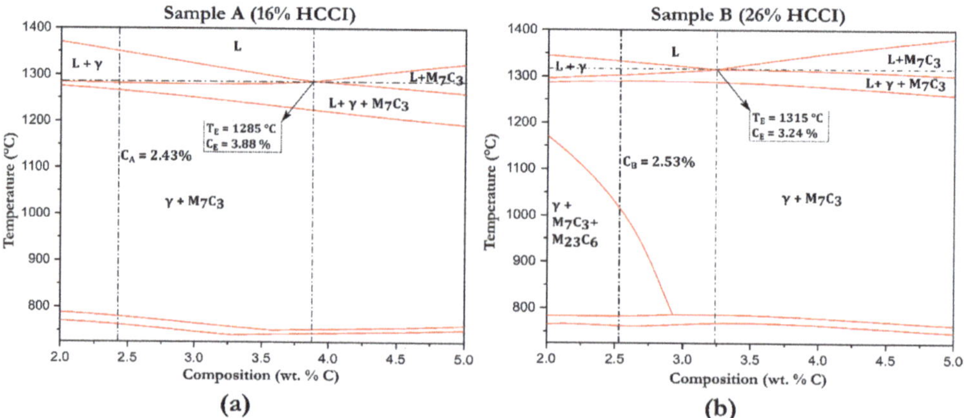

Figure 4. Pseudo-binary phase diagrams for samples, A and B computed using MatCalc 6. The eutectic transition temperature (T_E) and the corresponding eutectic carbon content (C_E) is indicated in the graphs along with the overall carbon content of (**a**) Sample A (C_A) and (**b**) Sample B (C_B).

Furthermore, to illustrate the influence of C and more specifically, Mo on the stability of the $M_{23}C_6$ carbide for various temperatures, phase boundaries were traced using MatCalc which is represented in Figure 5. It is observed, that at least 1% Mo is necessary for the low temperature stabilization of the $M_{23}C_6$ carbide at lower chromium contents for a constant carbon content, corroborating with other studies [10,41]. Moreover, at a given temperature, increasing the carbon content will necessitate an increase in the chromium content to ensure the stability of $M_{23}C_6$. This is because Cr is primarily a M_7C_3 carbide former for HCCI's up to 30% Cr after which, $M_{23}C_6$ becomes the stable carbide upon solidification [32,34].

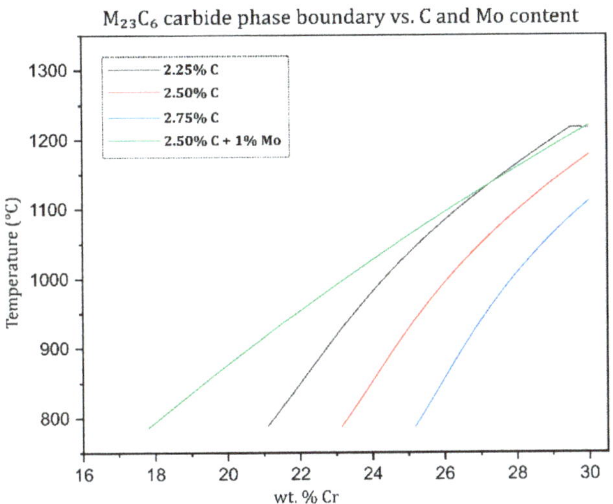

Figure 5. The influence of C and Mo on the phase boundary of the $M_{23}C_6$ carbide predicted by MatCalc 6.

3.3. Carbide Volume Fraction

It is well known that the nature and volume fraction of the EC is strongly influenced by the amount of chromium present in the alloy, as it is a strong carbide former [42]. The volume fraction of the EC corresponding to both samples, A and B was theoretically determined using empirical formulae suggested by Maratray et al. [29] (Equation (1)) and Doğan et al. [30] (Equation (2)), which was solely based on the bulk composition of C and Cr (in wt. %). Additionally, the total bulk chemical composition (Table 1) was used as the input for the software in simulating the equilibrium fraction of each phase. Figure 6 represents the equilibrium phase fractions determined for both alloys using MatCalc. The theoretical (Equations (1) and (2)) and simulated (MatCalc) values were then compared to the experimental results obtained from the I-A of microscopy images.

$$\% \text{ CVF} = 12.33\,(\%C) + 0.55\,(\%Cr) - 15.2 \tag{1}$$

$$\% \text{ CVF} = 14.05\,(\%C) + 0.43\,(\%Cr) - 22 \tag{2}$$

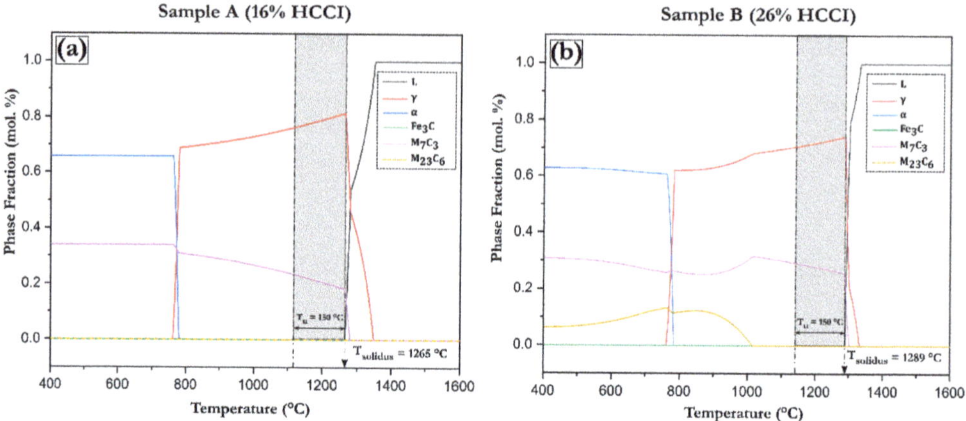

Figure 6. Equilibrium phase fractions for (**a**) Sample A (16% HCCI) and (**b**) Sample B (26% HCCI) as predicted by MatCalc. The solidus temperature, $T_{solidus}$ for both alloys including the range of undercooling considered, $T_u = 150\,^{\circ}C$ is indicated in the schematic.

Table 3 compares the % CVF obtained using the different approaches. It is evident that the % CVF of Sample B is higher than Sample A owing to the increased chromium for a quasi-constant carbon content. The % CVF increased by approximately 50% for an additional 10 wt. % of chromium. The differences in the results shown by the two formulae could be due to the number of samples considered for the study and the accuracy of determining the % CVF for each alloy [29,30]. Equation (1) was deduced by Maratray et al. after studying over 40 different alloys with varying C (1.95 to 4.31 wt. %) and Cr (10.8 to 25.82 wt. %) contents. Equation (2) is a result of the work carried out by Doğan et al. on hypoeutectic, eutectic and hypereutectic cast iron compositions with Cr contents of 15 and 26 wt. %. The hypoeutectic compositions considered in [30] is analogous to the alloys used in this study. Although the experimental values are close to the numerically predicted values, care must be taken as the formulae only consider the influence of C and Cr [1].

Table 3. Comparison between the % carbide volume fractions (CVFs) of both alloys obtained using different approaches.

Sample	Emperical Formulae (%)		MatCalc (at $T_{solidus}$) (%)	MatCalc (at $T_{solidus} - 150\,^{\circ}C$) (%)	Experimental (I-A) (%)
	Maratray [29]	**Doğan [30]**			
A (16% HCCI)	23.4 ± 2.1	19.0	18.4	21.2 ± 1.5	19.4 ± 0.4
B (26% HCCI)	30.6 ± 2.1	25.0	25.8	27.7 ± 1.1	30.4 ± 0.6

The equilibrium EC phase fraction predicted by MatCalc was initially considered at the temperature where the liquid ceases to exist i.e., the solidus temperature, $T_{solidus}$. Due to the larger solidification range of Sample A, its $T_{solidus}$ is lower (1265 °C) than Sample B (1289 °C). In both cases, the % CVF predicted by MatCalc is lower than the values obtained experimentally which can be explained by the non-equilibrium solidification during casting, leading to the existence of an undercooling regime [31,39]. Considering a degree of undercooling, $T_u = 150\,^{\circ}C$, and averaging the phase fraction values in that range ($T_{solidus} - 150\,^{\circ}C$) for both alloys yields a value of 21.2% ± 1.5% CVF for Sample A and 27.7% ± 1.1% CVF for Sample B. Comparing this % CVF with the value determined at $T_{solidus}$, an increase is observed in both cases (Sample A and Sample B). The increase in the % CVF with decreasing temperature can be attributed to the increased driving force for nucleation due to the undercooling effect. Consequently, it is observed that the predicted and experimental % CVF values correspond well and differences between them, fall within an error of less than 5%.

To further validate the results given by the simulation software, a few alloys with varying C and Cr contents were considered from the expansive study conducted by Maratray et al. [29] and Doğan et al. [30]. Table 4 lists the considered alloys and the % CVF obtained using metallography, Equations (1) and (2), and the prediction by MatCalc. The procedure employed to obtain the % CVF from MatCalc is similar to the one previously mentioned, i.e., determining the $T_{solidus}$ for each composition and considering an undercooling range, $T_u = 150\ °C$. It is seen that for relatively low Cr/C ratios (3.7–4.3), the formula suggested by Doğan and experimental values match, whereas Maratray's formula can be used for higher Cr/C ratios. It is also worth mentioning that increasing the Cr content alone did not lead to a significant increase in % CVF as evident in alloys M1 and M3. Comparing the alloys M1, D1 and M2, it is observed that addition of C (2.08 → 3.54 → 4.10) for a relatively constant Cr led to a massive increase in the % CVF (18.7 → 33.0 → 42.0). This further consolidates the effect of C on the volume fraction of the carbides compared with Cr. Despite the drastic variation in C and Cr content in all samples, the % CVF predicted by MatCalc is always comparable to the experimentally determined values which supports the usage of simulation to predict the volume fraction of the eutectic carbide in the as-cast state.

Table 4. The % CVF of certain alloys experimentally determined by [29,30], the numerical estimations and the corresponding % CVF computed using MatCalc (in grey).

Study	Alloy	Chemical Composition (wt. %)							Cr/C	% CVF (Experiment)	% CVF Equation (1)	% CVF Equation (2)	% CVF (MatCalc)
		C	Cr	Mn	Mo	Si	Ni	Fe					
Maratray et al. [29]	M1	2.08	15.85	0.70	-	1.00	-	Bal.	7.6	18.7	19.2 ± 2.1	14.0	16.8 ± 1.5
	M2	4.10	15.10	0.70	-	1.00	-	Bal.	3.7	42.0	43.7 ± 2.1	42.1	40.9 ± 1.2
	M3	2.08	20.55	0.70	-	1.00	-	Bal.	9.9	20.5	21.7 ± 2.1	16.1	19.7 ± 1.3
	M4	2.95	25.82	0.70	0.02	1.00	-	Bal.	8.8	32.3	35.4 ± 2.1	30.6	32.3 ± 1.0
Doğan et al. [30]	D1	3.54	15.2	0.61	0.31	0.51	0.18	Bal.	4.3	33.0 ± 2.0	36.8 ± 2.1	34.3	34.0 ± 1.4
	D2	2.76	26.2	0.93	0.38	0.42	0.38	Bal.	9.5	29.0 ± 1.0	33.2 ± 2.1	28.0	30.4 ± 1.1

3.4. Chemical Composition

In order for effectively designing heat treatment cycles, it is essential to have an understanding of the individual matrix and carbide compositions [30]. The composition between the matrix and carbide will vary depending upon the alloying elements present, the cooling rate during casting and the bulk Cr/C ratio [1,5]. For these reasons, and in order to correlate experimental with simulated values, the matrix and carbide chemical composition were determined by EPMA and MatCalc (Table 5). The matrix and carbide elemental compositions determined with MatCalc, were calculated considering the weight fraction of the element present in the respective phase at the given temperature and taking the average value over the undercooling range ($T_{solidus} - T_u$).

Table 5. Matrix and carbide elemental compositions (in wt. %) for the two alloys as determined by electron probe microanalysis (EPMA) and MatCalc. The distribution of Cr and Fe within the M_7C_3 carbide is also compared.

Sample	Matrix			Carbide			$(Cr_xFe_y)C_3$	
	Element (wt. %)	EPMA	MatCalc	Element (wt. %)	EPMA	MatCalc	EPMA	MatCalc
A (16% HCCI)	C	0.86 ± 0.34	1.12 ± 0.12	C	7.54 ± 0.49	8.71 ± 0.00	$(Cr_{3.9}Fe_{3.1})C_3$	$(Cr_{3.9}Fe_{3.1})C_3$
	Cr	12.10 ± 0.21	8.79 ± 0.71	Cr	48.81 ± 3.63	49.5 ± 0.45		
	Mn	0.60 ± 0.02	0.77 ± 0.00	Mn	-	-		
	Ni	0.13 ± 0.03	0.21 ± 0.00	Ni	0	0		
	Mo	0.21 ± 0.04	0.27 ± 0.03	Mo	-	-		
	Fe	85.6 ± 0.3	88.84 ±0.85	Fe	41.99 ± 3.65	39.99 ± 0.52		
B (26% HCCI)	C	0.43 ± 0.13	0.69 ± 0.09	C	7.85 ± 0.53	8.83±0.00	$(Cr_5Fe_2)C_3$	$(Cr_5Fe_2)C_3$
	Cr	18.21 ± 1.24	15.37 ± 0.78	Cr	63.07 ± 1.87	65.13 ± 0.77		
	Mn	0.67 ± 0.03	0.72 ± 0.00	Mn	-	-		
	Ni	0.20 ± 0.04	0.33 ± 0.00	Ni	0	0		
	Mo	0.14 ± 0.03	0.2 ± 0.01	Mo	-	-		
	Fe	80.0 ± 1.3	82.68 ± 0.87	Fe	28.36 ± 2.11	25.19 ± 0.79		

From Table 5, it is evident that the Cr content (wt. %) of both the matrix and EC increased as the bulk Cr content increased although the C content (wt. %) in the EC remained at a stoichiometric level. The addition of chromium reduces the carbon solubility in austenite [17] and as a result, the 26% HCCI alloy has a lower matrix C content. This can be further elucidated by considering the partition ratio of the elements. The segregation or partition ratio, as coined by Laird [43], is the element's affinity to partition into the carbide or the matrix and can be defined as the ratio of the weight percent of the element in the carbide to the matrix. Higher ratios, as in the case of chromium and carbon suggest strong partitioning towards the carbides whereas elements such as Si, Ni, and Cu are found only at the matrix regions. The Cr partition ratio from EPMA measurements for Sample A and Sample B was 4 and 3.5, respectively. Furthermore, the partition ratio of C in Sample B (18.25) is higher compared to Sample A (8.77) indicating an increased affinity to the eutectic carbides (as evidenced by the lower C content of the 26% HCCI matrix).

The distribution of Cr and Fe within the EC was determined by converting the respective weight percentages into atomic percentage and normalizing with the carbon atomic percentage (30 at. %). Despite both materials having M_7C_3 as the EC, the Fe/Cr ratio is lower in the case of 26% HCCI indicating that less Cr atoms were substituted by Fe. It is also worthy to mention that although trace amounts of Mn and Mo were detected by EPMA and also predicted by MatCalc in the EC for both samples, it is not shown in Table 5. The Cr content (in at. %) of the EC increased from approximately 40% in Sample A to 50% in Sample B, which is the highest for these type of alloys [29]. In both cases, the elemental compositions of the matrix and EC, and the chromium–iron distribution in the EC predicted by MatCalc are in accordance with the values measured by EPMA.

A similar alloy (C, 2.72%; Cr, 26.6%; Mn, 0.2%; Si, 0.78%; Ni, 0.17%) was studied by Carpenter et al. [44] wherein the Cr content of the EC was determined to be 49.7 at. % ± 1.6 at. % by chemical microanalysis. Comparing this alloy with Sample B, it is observed that the Cr content of the EC is identical in both cases even though there is a slight increase in the bulk C content. This further upholds the notion that 50 at. % Cr ($Cr_5Fe_2C_3$) is the highest for M_7C_3 type of EC in HCCIs [33]. Moreover, the experimental value obtained by Carpenter et al. was corroborated by MatCalc, which predicted a Cr content of 49.7 at. % ± 0.7 at. % Cr in the EC.

3.5. Hardness

Table 6 details the values of hardness on three different scales. Sample B shows higher bulk hardness (HRC) owing to the higher volume fraction of the M_7C_3 carbides formed. Despite the differences in the chemical composition of the matrix, its microhardness (HV0.1) for both alloys remained similar. This coincidental value of the matrix hardness can be attributed to the presence of a high carbon matrix in Sample A and a high alloying in Sample B due to the Cr content [30]. It also sheds light on the efficacy of the carbon contribution to the hardness of the matrix compared to chromium. Furthermore, the hardness of the M_7C_3 carbide increases from Sample A to Sample B which could be attributed to the increasing Cr content of the carbide [45].

Table 6. Results of the bulk, matrix and carbide hardness for both alloys.

Sample	Rockwell (HRC) (Bulk)	Vickers (HV0.1) (Matrix)	Nanoindentation (GPa) (EC)
A (16% HCCI)	46.3 ± 0.8	356 ± 11	13.1 ± 1.7
B (26% HCCI)	49.3 ± 0.5	360 ± 21	19.0 ± 1.2

4. Conclusions

Computational tools and experimental results were combined in this work with the aim to validate the usage of MatCalc simulations for the prediction of phase fractions and elemental distribution in HCCI's in the as-cast condition, for the convincing implementation of these tools for further heat treatment design. For that, two as-cast alloys (containing 16% and 26% Cr) were fabricated under similar conditions with the main variance being the Cr content. The EC phase fraction and the individual matrix, and carbide compositions were determined experimentally and compared with the values predicted by the MatCalc simulation. The principal conclusions that can be drawn from the above work are as follows:

- OM and SEM micrographs indicate a dispersion of M_7C_3 eutectic carbides in an austenitic matrix with a thin layer of martensite formed at the carbides' periphery for both alloys. Although the matrix is purely austenite in Sample B, some partial transformation to pearlite has occurred in Sample A, owing to the low Cr/C ratio.
- The pseudo-binary phase diagrams constructed using MatCalc indicate the formation of $M_{23}C_6$ carbide at temperatures below 1050 °C in Sample B. However, the presence of $M_{23}C_6$ carbide was not detected due to the non-equilibrium cooling and the low Mo content of the alloy.
- An addition of 10 wt. % Cr lead to an increase of about 50% of the EC as evidenced by image analysis. The % CVF was lower when determined from MatCalc at $T_{solidus}$. Nevertheless, by considering an undercooling range of 150 °C as a consequence of the non-equilibrium solidification, the predicted % CVF corresponded well with experimentally determined values. Therefore, the MatCalc simulated data is reliable for the determination of % CVF. The accuracy of the simulation software was further validated comparing the % CVF of several alloys (with different C and Cr contents) to the experimental values obtained by other authors from metallographic techniques.
- The predictions made by MatCalc are in accordance with the values obtained by EPMA. MatCalc also predicted an increase Cr/Fe ratio in the EC with increasing Cr content, which was corroborated by EPMA measurements. Additionally, the Cr/Fe ratio predicted by MatCalc for Sample B showed a good correspondence with experimental results found in the literature.
- Finally, the increase in the bulk hardness of Sample B was related to the increased M_7C_3 fraction, whereas the individual EC hardness was higher in Sample B than Sample A due to the increased Cr occupation in the EC. Despite the lower bulk and carbide hardness, the matrix hardness of Sample A was on par with B, probably due to the high C content in the matrix which prevented a decrement.

To sum up, this work demonstrated the capability of MatCalc to accurately predict the EC phase fraction and elemental distribution within the phases, which bolsters its implementation in the design of heat treatments. The time and resource intensive experimental procedures can be replaced by simulation techniques to determine the phase fraction and especially, the individual phase compositions in the as-cast state. Furthermore, the elemental distribution within the matrix and EC is reflected in its corresponding hardness. The knowledge provided by this tool about the elemental distribution within the phases beforehand will assist in designing a heat treatment cycle for an HCCI alloy to be used in a specific application and pave way for 'microstructural tailoring'. Accordingly, the microstructural modifications occurring in these alloys during heat treatments, including the carbide precipitation kinetics, will be investigated as a part of future work.

Author Contributions: Conceptualization and methodology, U.P.N. and M.A.G.; Simulation, experimental validation, formal analysis, data curation and writing–original draft preparation, U.P.N.; writing–review and editing, and supervision, M.A.G. and F.M.; project administration and funding acquisition, F.M. All authors have read and agreed to the published version of the manuscript.

Funding: The authors wish to acknowledge the EFRE Funds (C/4-EFRE-13/2009/Br) of the European Commission for supporting activities within the AME-Lab project.

Acknowledgments: The authors would like to thank Martin Duarte from Tubacero S.A. for providing the materials and Sandvik Materials Technology for the EPMA measurements. Additionally, UP Nayak is grateful to DAAD for the financial support.

Conflicts of Interest: The authors declare no conflict of interest.

References

1. Tabrett, C.P.; Sare, I.R.; Ghomashchi, M.R. Microstructure-property relationships in high chromium white iron alloys. *Int. Mater. Rev.* **1996**, *41*, 59–82. [CrossRef]
2. ASTM Standard A 532/A 532M-93a. *Standard Specification for Abrasion Resistance Cast Irons*; ASTM International: West Conshohocken, PA, USA, 2003. Available online: www.astm.org (accessed on 23 December 2019).
3. International Organization for Standardization. *Abrasion-Resistant Cast Irons-Classification (ISO 21988:2006)*; International Organization for Standardization: Geneva, Switzerland, 2006. Available online: www.iso.org (accessed on 23 December 2019).
4. Karantzalis, E.; Lekatou, A.; Mavros, H. Microstructure and properties of high chromium cast irons: Effect of heat treatments and alloying additions. *Int. J. Cast Met. Res.* **2009**, *22*, 448–456. [CrossRef]
5. Wiengmoon, A.; Pearce, J.T.H.; Chairuangsri, T. Relationship between microstructure, hardness and corrosion resistance in 20 wt. %Cr, 27 wt. %Cr and 36 wt. %Cr high chromium cast irons. *Mater. Chem. Phys.* **2011**, *125*, 739–748. [CrossRef]
6. Laird, G.; Powell, G.L.F. Solidification and solid-state transformation mechanisms in Si alloyed high-chromium white cast irons. *Metall. Trans. A* **1993**, *24*, 981–988. [CrossRef]
7. Bedolla-Jacuinde, A.; Correa, R.; Quezada, J.G.; Maldonado, C. Effect of titanium on the as-cast microstructure of a 16%chromium white iron. *Mater. Sci. Eng. A* **2005**, *398*, 297–308. [CrossRef]
8. Kopyciński, D.; Guzik, E.; Siekaniec, D.; Szczęsny, A. The effect of addition of titanium on the structure and properties of high chromium cast iron. *Arch. Foundry Eng.* **2015**, *15*, 35–38. [CrossRef]
9. Scandian, C.; Boher, C.; de Mello, J.D.B.; Rézaï-Aria, F. Effect of molybdenum and chromium contents in sliding wear of high-chromium white cast iron: The relationship between microstructure and wear. *Wear* **2009**, *267*, 401–408. [CrossRef]
10. Imurai, S.; Thanachayanont, C.; Pearce, J.T.H.; Tsuda, K.; Chairuangsri, T. Effects of Mo on microstructure of as-cast 28 wt. % Cr-2.6 wt. % C-(0-10) wt. % Mo irons. *Mater. Charact.* **2014**, *90*, 99–112. [CrossRef]
11. Lv, Y.; Sun, Y.; Zhao, J.; Yu, G.; Shen, J.; Hu, S. Effect of tungsten on microstructure and properties of high chromium cast iron. *Mater. Des.* **2012**, *39*, 303–308. [CrossRef]
12. Cortés-Carrillo, E.; Bedolla-Jacuinde, A.; Mejía, I.; Zepeda, C.M.; Zuno-Silva, J.; Guerra-Lopez, F.V. Effects of tungsten on the microstructure and on the abrasive wear behavior of a high-chromium white iron. *Wear* **2017**, *376–377*, 77–85. [CrossRef]
13. Doğan, Ö.N.; Hawk, J.A. Effect of carbide orientation on abrasion of high Cr white cast iron. *Wear* **1995**, *189*, 136–142. [CrossRef]
14. Gahr, K.H.Z.; Doane, D.V. Optimizing fracture toughness and abrasion resistance in white cast irons. *Metall. Trans. A* **1980**, *11*, 613–620. [CrossRef]
15. Asensio, J.; Pero-Sanz, J.A.; Verdeja, J.I. Microstructure selection criteria for cast irons with more than 10 wt. % chromium for wear applications. *Mater. Charact.* **2002**, *49*, 83–93. [CrossRef]
16. Maratray, F.; Poulalion, A. Austenite Retention in High-Chromium White Irons. *Trans. Am. Foundrym. Soc.* **1982**, *90*, 795–804.
17. Poolthong, N.; Nomura, H.; Takita, M. Effect of heat treatment on microstructure and properties of semi-solid chromium cast iron. *Mater. Trans.* **2004**, *45*, 880–887. [CrossRef]
18. *ASM Handbook Volume 3: Alloy Phase Diagrams*; ASM International: Materials Park, OH, USA, 2016; ISBN 978-1-62708-070-5. Available online: www.asminternational.org (accessed on 23 December 2019).
19. Kozeschnik, E.; Buchmayr, B. Matcalc—A simulation tool for multicomponent thermodynamics, diffusion and phase transformations. In Proceedings of the Mathematical Modelling of Weld Phenomena, Leibnitz, Austria, 1–3 October 2001; pp. 349–361.
20. Kroupa, A. Modelling of phase diagrams and thermodynamic properties using Calphad method—Development of thermodynamic databases. *Comput. Mater. Sci.* **2013**, *66*, 3–13. [CrossRef]
21. Li, D.; Liu, L.; Zhang, Y.; Ye, C.; Ren, X.; Yang, Y.; Yang, Q. Phase diagram calculation of high chromium cast irons and influence of its chemical composition. *Mater. Des.* **2009**, *30*, 340–345. [CrossRef]
22. Albertin, E.; Beneduce, F.; Matsumoto, M.; Teixeira, I. Optimizing heat treatment and wear resistance of high chromium cast irons using computational thermodynamics. *Wear* **2011**, *271*, 1813–1818. [CrossRef]
23. Akyildiz, Ö.; Candemir, D.; Yildirim, H. Simulation of Phase Equilibria in High Chromium White Cast Irons. *Uludağ Univ. J. Fac. Eng.* **2018**, *23*, 179–190. [CrossRef]

24. Guitar, M.A.; Scheid, A.; Suárez, S.; Britz, D.; Guigou, M.D.; Mücklich, F. Secondary carbides in high chromium cast irons: An alternative approach to their morphological and spatial distribution characterization. *Mater. Charact.* **2018**, *144*, 621–630. [CrossRef]

25. Guitar, M.A.; Suárez, S.; Prat, O.; Duarte Guigou, M.; Gari, V.; Pereira, G.; Mücklich, F. High Chromium Cast Irons: Destabilized-Subcritical Secondary Carbide Precipitation and Its Effect on Hardness and Wear Properties. *J. Mater. Eng. Perform.* **2018**, *27*, 3877–3885. [CrossRef]

26. Collins, W.K.; Watson, J.C. Metallographic etching for carbide volume fraction of high-chromium white cast irons. *Mater. Charact.* **1990**, *24*, 379–386. [CrossRef]

27. Wiengmoon, A.; Chairuangsri, T.; Poolthong, N.; Pearce, J.T.H. Electron microscopy and hardness study of a semi-solid processed 27 wt%Cr cast iron. *Mater. Sci. Eng. A* **2008**, *480*, 333–341. [CrossRef]

28. Schindelin, J.; Ignacio Arganda-Carreras, E.F.; Kaynig, V.; Longair, M.; Pietzsch, T.; Preibisch, S.; Rueden, C.; Saalfeld, S.; Schmid, B.; Tinevez, J.-Y.; et al. Fiji: An open-source platform for biological-image analysis. *Nat. Methods* **2012**, *9*, 676–682. [CrossRef]

29. Maratray, F.; Usseglio-Nanot, R. *Factors Affecting the Structure of Chromium and Chromium—Molybdenum White Irons*; Climax Molybdenum S. A: Paris, France, 1970.

30. Doğan, Ö.N.; Hawk, J.A.; Laird, G. Solidification structure and abrasion resistance of high chromium white irons. *Metall. Mater. Trans. A Phys. Metall. Mater. Sci.* **1997**, *28*, 1315–1328. [CrossRef]

31. Bedolla-Jacuinde, A.; Hernández, B.; Béjar-Gómez, L. SEM study on the M7C3 carbide nucleation during eutectic solidification of high chromium white irons. *Zeitschrift Met.* **2005**, *96*, 1380–1385.

32. Zumelzu, E.; Goyos, I.; Cabezas, C.; Opitz, O.; Parada, A. Wear and corrosion behaviour of high-chromium (14–30% Cr) cast iron alloys. *J. Mater. Process. Technol.* **2002**, *128*, 250–255. [CrossRef]

33. Maratray, F. Choice of appropriate compositions for chromium-molybdenum white irons. *AFS Trans.* **1971**, *79*, 121–124.

34. Abdel-Aziz, K.; El-Shennawy, M.; Omar, A.A. Microstructural Characteristics and Mechanical Properties of Heat Treated High-Cr White Cast Iron Alloys. *Int. J. Apll. Eng. Res.* **2017**, *12*, 4675–4686.

35. Doğan, Ö.N.; Laird, G.; Hawk, J.A. Abrasion resistance of the columnar zone in high Cr white cast irons. *Wear* **1995**, *181*, 342–349. [CrossRef]

36. Matsubara, Y.; Ogi, K.; Matsuda, K. Eutectic Solidification of High-Chromium Cast Iron—Eutectic Structures and Their Quantitative Analysis. *Trans. Am. Foundrym. Soc.* **1982**, *89*, 183–196.

37. Ohide, T.; Ohira, G. Solidification of High Chromium Alloyed Cast Iron. *Foundryman* **1983**, *76*, 7–14.

38. Doğan, Ö.N. Columnar to equiaxed transition in high Cr white iron castings. *Scr. Mater.* **1996**, *35*, 163–168. [CrossRef]

39. Filipovic, M.; Romhanji, E.; Kamberovic, Z. Chemical composition and morphology of M7C3 eutectic carbide in high chromium white cast iron alloyed with vanadium. *ISIJ Int.* **2012**, *52*, 2200–2204. [CrossRef]

40. Medvedeva, N.I.; Van Aken, D.C.; Medvedeva, J.E. Stability of binary and ternary M23C6 carbides from first principles. *Comput. Mater. Sci.* **2015**, *96*, 159–164. [CrossRef]

41. Choi, J.W.; Chang, S.K. Effects of Molybdenum and Copper Additions on Microstructure of High Chromium Cast Iron Rolls. *ISIJ Int.* **1992**, *32*, 1170–1176. [CrossRef]

42. Powell, G.L.F.; Laird, G. Structure, nucleation, growth and morphology of secondary carbides in high chromium and Cr-Ni white cast irons. *J. Mater. Sci.* **1992**, *27*, 29–35. [CrossRef]

43. Laird, G. Microstructures of Ni-Hard I, Ni-Hard IV and High-Cr White Cast Irons. *AFS Trans.* **1991**, *99*, 339–357.

44. Carpenter, S.D.; Carpenter, D.; Pearce, J.T.H. XRD and electron microscope study of an as-cast 26.6% chromium white iron microstructure. *Mater. Chem. Phys.* **2004**, *85*, 32–40. [CrossRef]

45. Kagawa, A.; Okamoto, T.; Saito, K.; Ohta, M. Hot hardness of (Fe, Cr)3C and (Fe, Cr)7C3 carbides. *J. Mater. Sci.* **1984**, *19*, 2546–2554. [CrossRef]

MDPI

St. Alban-Anlage 66

4052 Basel

Switzerland

Tel. +41 61 683 77 34

Fax +41 61 302 89 18

www.mdpi.com

Metals Editorial Office

E-mail: metals@mdpi.com

www.mdpi.com/journal/metals